T0366676

THE KILLER WHALE JOURNALS

Hanne Strager

Foreword and Photographs by Paul Nicklen

THE KILLER WHALE JOURNALS

OUR LOVE AND FEAR OF ORCAS

JOHNS HOPKINS UNIVERSITY PRESS
Baltimore

Printed in the United States of America on acid-free paper
9 8 7 6 5 4 3 2 1

Johns Hopkins University Press
2715 North Charles Street
Baltimore, Maryland 21218
www.press.jhu.edu

Library of Congress Cataloging-in-Publication Data

Names: Strager, Hanne, author.
Title: The killer whale journals : our love and fear of orcas / Hanne Strager ;
foreword and photographs by Paul Nicklen.
Description: Baltimore : Johns Hopkins University Press, [2023] | Includes
bibliographical references and index.
Identifiers: LCCN 2022023264 | ISBN 9781421446226 (hardcover ; acid-free paper) |
ISBN 9781421446233 (ebook)
Subjects: LCSH: Killer whale. | Killer whale—Behavior. | Predatory animals.
Classification: LCC QL737.C432 S79 2023 | DDC 599.53/6—dc23/eng/20220630
LC record available at https://lccn.loc.gov/2022023264

A catalog record for this book is available from the British Library.

In the gallery that follows page 98, photographs by Paul Nicklen appear
on pages 1, 5, 6, 7, and 8.

To
Svea and Alva

and to the memory
of Per Ole Lund

Contents

Foreword

When I was a young boy, I would dream about being close to animals. All animals. Any animal would do. I had a pet seagull that I adored, even if others around me didn't care much for seagulls. Large predators, though, were my favorites. Their majestic bearing, their looming size, and their enigmatic, often unpredictable behavior piqued my interest and left me enthralled. I imagined what it would be like to be part of their world.

I am no longer a boy, though part of that young child remains. As an adult, I have been fortunate to fulfill my childhood dream and bring the underwater world of leopard seals, sharks, polar bears, and orcas to the surface, through photography. As an adult, I am now privileged to share my experiences with others.

Predators are often misunderstood. They are demonized and feared in equal measure. Orcas are no exemption. Just look at their more familiar, better-known nickname: killer whales. They are not whales—they are more closely related to the dolphin family—and they kill only for food, not for sport.

Orcas are seen as relentless, merciless hunters. They have been persecuted for their perceived cunning and inherently evil nature. Today, we know that this is nonsense. Not only do orcas breathe air like us, they love, they mourn, they play, and they communicate among themselves, much as we do. There are few life moments on this planet that beat the experience of meeting a large orca underwater, in its own environment, seeing eye to eye.

Years ago, during an expedition to Norway, I was in the water with a pod of orcas. They were feeding on a large school of herring, as is their habit, when a particularly large male decided to check me out. He approached slowly until he was just a few feet away, a small dark eye inspecting me out of curiosity as he swam slowly by. He passed my diving companion Göran Ehlmé, equally curious, then turned and approached me again. To feel the curiosity of another being and to connect with that being on such a fundamental and yet serendipitous level was both a humbling and blissful experience.

I first met Hanne in the field, too, in northern Norway. We would go out every day to document and film the orcas native to that area. I soon learned that we shared a mutual love and understanding for everything wild. Hanne was—and remains—truly passionate about orcas.

It is one of the greatest gratifications of my life that orcas have found such a strong voice through Hanne's writing. *The Killer Whale Journals* is a page-turner in every sense of the word. On one level, it's a reasoned account of humankind's often troubled relationship with orcas over the years and decades. On another, more fundamental level, it's an honest, eye-opening examination of our often extraordinary relationship with orcas over the years and decades.

These magnificent marine predators are still met with distrust and suspicion in some instances, if not with outright hate. But they are also at the heart of incredible collaborative relationships with people, from marine biologists to oceanographers to conservation photographers, such as myself.

Hanne's book is as much about people as it is about orcas. That's important. How we as humans choose to live with the natural world's apex predators will define who we are to future generations. Were we hunters who merely wanted to subjugate them and lead them to ultimate extinction, or were we seeking a way to coexist in peace? Ultimately, what we decide will not only shape the world the orcas live in but the world we ourselves will be living in.

It is my fervent hope you will enjoy this book as much as I did.

Paul Nicklen

THE KILLER WHALE JOURNALS

Prologue

It was just past midday and already the light was low. In the Lofoten Islands in northern Norway, the sun only just climbs above the horizon in October. Earlier in the morning, I had arrived to Svolvær after almost two days of travel from Denmark. In Svolvær, a local bus took me to my destination, the small fishing village of Henningsvær, a bit further down the road where I had volunteered to be the cook on a small converted fishing vessel. I had no previous experience with either cooking or fishing vessels, but I had been brought up on Pippi Longstocking and had adopted her motto, "I've never tried that before, so I think I should definitely be able to do that."

I knew almost nothing about the boat or the people aboard it before I arrived. A few days earlier, I had been standing in line behind a tall dark-haired guy wearing an Icelandic wool sweater at the cafeteria at Aarhus University in Denmark, where I was in my third year studying biology. He studied biology as well but was a few years ahead of me. To pass the time we started chatting, and he introduced himself as Morten Lindhard, telling me that he was about to go on a trip looking for killer whales in Norway. It sounded like everything I had secretly dreamed of—I had read Jane Goodall's *In the Shadow of Man* and seen the documentaries about her work. I had been captivated by the ocean films and underwater footage that Jacques Cousteau had produced from his many years on board the research vessel *Calypso*. I loved studying biology, but I was hungry for adventure.

"Is there any way I could come along?" I asked brazenly. It turned out that they were still looking for a cook, so I called up the skipper to see if they would take me on. My interview consisted of two questions: Could I cook? I said yes. Did I get seasick? I said no. I immediately got the job and, with two days' notice, I crammed my backpack with woolen sweaters and bought a cookbook and a new pair of rubber boots. I was still in the middle of a semester, but at that time, it was relatively easy to grant yourself a leave of absence from the university as long as you showed up for the exams and mandatory lab classes. I did make it back in time to take my exams at the end of the semester, and for the mandatory lab classes I was fortunate to have an identical twin sister who volunteered to sign in for me.

Old-Bi was a light blue, wooden fishing boat, built in Denmark in the 1950s. The skipper had converted the 20-meter cutter into a liveaboard for researchers and conservationists and named it after his grandmother. It was no problem finding *Old-Bi* in the harbor when I arrived in Henningsvær. With wetsuits drying on the railing, waterproof camera cases stacked on the deck, and people crawling all over the pier and boat, it easily stood out among the more traditional-looking fishing boats. I had arrived at a hectic moment. The skipper had just been called on the VHF radio by a local fisherman who had spotted killer whales somewhere outside the harbor. Everybody was getting ready to board the two small inflatable Zodiacs that were tied to the side of *Old-Bi*. From their conversations, I understood that it was something of a breakthrough that the call on the VHF came from a fisherman. When I asked why, they replied, "Many people here don't like killer whales very much."

The skipper gave me a hand climbing on board *Old-Bi* with my luggage and then asked if I wanted to come along in one of the Zodiacs. "It's now," he said. "We are leaving as soon as possible. The whales move fast so we need to get out there." Without hesitation, I agreed. I hurried to get dressed in as many layers as possible and dug out my new rubber boots from my backpack before I clumsily dumped myself into one of the inflatables. As I sat on the pontoon holding onto a rope that ran alongside the boat, I started to feel a bit nervous. I had never been in a Zodiac before and the thrill of sitting this close to the sea and seeing killer whales made my heart beat fast. There were three men and another woman in the boat. I knew Morten from the meeting in the cafeteria but had not met the others before. There was no time for introductions, though, before we zoomed out of the harbor at high speed.

Once we were out of the protected waters, the waves grew bigger. We were roller-coasting up and down the blue peaks and troughs and could only scan the horizon for whales when we were at the top of a wave. While trying to hold onto the rope with my drenched woolen mittens, I was also anxious not to show how inexperienced I was. And nervous about our safety. I pondered both the big waves and the killer whales and decided that since nobody else looked especially worried it was probably ok. They were all concentrated on looking for the whales.

I wasn't sure what to look for, and so I asked one of the guys sitting opposite me for some pointers. "Well, usually we don't find them," he laughed. He then added, "But it is a good idea to look out for seagulls. If there are many seagulls flocking in an area it sometimes means that there are killer whales there too."

The driver of the Zodiac tried to maneuver as best as he could but every now and then he hit a wave just so and water cascaded onto us.

Now I understood why everybody else was dressed in one-piece watertight survival suits. My clothes were soaked, and frigid seawater filled my boots. But soon enough we were in the area where the fisherman had seen the whales.

To the north and west of us, the sea was packed with skerries and small rocky islands. To the south and east was the open sea. The woman next to me pointed to some seagulls close to a low rocky island with grass on top. "Could that be it?" she asked. The driver nodded and changed direction, heading straight for the grassy island. I squinted my eyes, but I saw only water. Then suddenly, out of nowhere, the killer whales appeared, not in front of us but behind us.

I saw only a quick glimpse of several black fins rising out of the water before they disappeared again. The driver immediately slowed down. Even over the noise from the outboard engine, I heard the powerful exhalations from their blowholes as they came up for air. At first, I thought they were moving in the same direction as us. Then it dawned on me that they were not only moving in the same direction, but they were moving toward us, following us in the wake of the boat. My heart was racing with excitement—and maybe just a little alarm. The guy sitting opposite me must have sensed my apprehension, because he leaned over and whispered, "It's ok, they do this."

I counted at least six to eight animals of different sizes. From the little reading I had managed to do before setting off, I could tell that the ones with towering dorsal fins were males and the others with much smaller fins were either females or juveniles. The whole group came up in unison a few times and then disappeared below the surface again. We continued on the same course for a few minutes, and once again the whales surfaced. Much closer this time. They were now swimming next to the boat. Their fins cut through the water as their massive black bodies broke the surface. Their exhalations sounded like explosions and the mist of their blows hung in the air. My heart was still pounding, though more from the exhilaration of being so close to one of the biggest predators on Earth than from fear. I sensed that they were not threatening; they were just coming along for the ride.

A large male came up right next to the boat, so close that I could see water running down his gleaming skin. A pearly black eye just in front of the white eyepatch stared right at me. It was just a quick moment, but it stayed with me after the whale was gone. I realized that this huge killer whale had been checking us out—just as we were checking them out. To sense the awareness and curiosity of another being, and perhaps even its desire to connect, shatters an invisible barrier. It perforates the solitude of being human in a wild world where we are surrounded by creatures we don't understand and can't reach.

The trip back was cold and wet, but it didn't matter. I was euphoric and the view was breathtaking. The Lofoten Islands are a chain of mountainous islands that stretch from the Norwegian mainland hundreds of kilometers into the North Atlantic Ocean. The serrated peaks rise almost vertically from sea level to some 1,000 meters of elevation. Below the nearest massive cliff was the entrance to the harbor.

In the days that followed, I did my best to step into the role of ship's cook. I filleted bucketfuls of gleaming herring and marinated them in sour cream and garlic while being tossed around in heavy seas in a galley that filled with diesel fumes. Much to my own surprise, I learned that I was quite sturdy at sea, and I began to think that I could grow into this role. During those early days I also learned more about the project I had become a part of.

Our mission was to find a suitable place in Norway to establish whale-watching and to combine tourism with whale research. The people on board were a lively assortment of individuals. The skipper, Mic Calvert, was a photographer from Gothenburg in Sweden; his girlfriend, Bibbi Forsman, an artist; others were journalists or entrepreneurs; and a few were biologists or students like me. I felt incredibly lucky. Studying biology at university was very theoretical, and I had not yet found out what my own direction should be. Being on a boat surrounded by wilderness was definitely closer to a possible path (even at the price of being up to my elbows in fish guts).

The Lofoten Islands had been chosen because the area was reputed to be a good place to find killer whales, especially in the fall and winter. We were there in October, so the season was right, but fall and winter in northern Norway are also the time for strong winds and stormy weather. Additionally, north of the Arctic Circle the days are short as wintertime approaches and the daylight hours are few. The plan was to use *Old-Bi* as our base and to go out searching for whales using the small inflatables we were towing behind her. Most days the crews came back cold and wet having seen nothing except for the spectacular landscape of the ragged mountains adorned with low hanging clouds. A few times they actually found killer whales, but it wasn't easy to catch up with them or to follow them out in the open sea.

Old-Bi was moored in the harbor of Henningsvær, which in turn was sheltered among an archipelago of islets and skerries on the southernmost edge of a larger island in the middle of the chain. All around the village, bedecked with small brightly painted wooden houses, we saw wooden racks for drying cod. The houses were huddled together on the few scraps of level ground found on the island, yet despite the lack of land, each had a tiny garden, fenced in to protect the windblown bushes and crippled trees inside.

The locals observed us with some suspicion. To them we were undoubtedly a bunch of hippies (depending on the definition, this was mostly true) or, even worse, affiliated with Greenpeace (also true for some of the people aboard). The latter was a graver sin because Norway was (and still is) a whaling country, and many Norwegians, especially in the north, felt strongly that this was not something people from the outside should interfere with. The International Whaling Commission had decided on a worldwide moratorium on whaling just a few years earlier (in 1982). This decision provoked immense debate in Norway, and the Norwegian government struggled with it until they decided to file an objection, freeing themselves from being obligated to abide by the moratorium. The debate was perhaps not as much about whaling as a business and livelihood as it was about the Norwegians sense of pride and independence. Many felt that nobody from the outside should tell Norwegians what was the right thing to do in their own country.

I arrived in Norway relatively ignorant of this explosive situation, and the attitudes of the people on board *Old-Bi* were mixed. Some were motivated by the possibility of establishing an alternative to whaling by creating an economic alternative based on whale-watching. For others their main interest was in the scientific potential of studying a species about which almost nothing was known in Norway. The expedition itself had been initiated by a small organization called the Center for Studies of Whales and Dolphins, which was based in Gothenburg, Sweden. We were a diverse group of Scandinavians with different backgrounds. As well as the biologists, there was a varying blend of journalists, artists, and photographers, of whom about eight to ten might have been on board at any one time.

Although we were in Norway, there was only one Norwegian involved. His name was Lars Øivind Knutsen, and he had just arrived from Svalbard where he had studied polar bears and walruses. Tall and blond with ruffled hair, narrow smiling eyes, and always wearing a home-knitted Norwegian sweater, he looked like an archetypical Scandinavian. The reason he was the only Norwegian was that it was difficult for the project to attract students and scientists from Norway. They had little interest in our work. On the contrary, established researchers didn't see it as a serious project. Despite Norway's incredible coastline (third longest in the world, surpassed only by Canada and Indonesia) and waters rich in all kind of marine life, the scientific interest in whales was narrow. A Norwegian biologist interested in marine mammals in the late 1980s generally worked with research connected to whaling, and the only whale species that the Norwegian authorities were interested in, therefore, was the minke whale. Lars Øivind's Norwegian colleagues saw his association with our "hippie-ship" as dubious, and it tainted him as a romantic, rather than a serious scientist.

Local townspeople and fishermen regarded us with a good deal of skepticism, too. But they were very friendly and sometimes they would tell us if they had seen killer whales. Every now and then, they would even give me a few buckets full of herring to marinate. And each morning they would wave as a couple of us from the expedition would clamber up the rocks behind the local school to a lookout we had set up on a small hilltop. Carrying folding chairs, a telescope with a tripod, and a backpack with a thermos and coffee mugs, we would station ourselves with a VHF radio to contact our dinghy crews and direct them in case we saw any killer whales.

The path behind the school twisted around the rocky outcrops and past the triangular wooden racks, which would later in the winter be used to dry the cod. It ended at the highest point of the island where we had a 360-degree view of the whole area. To call that view breathtaking is not really enough; it calls for stronger and more powerful adjectives. But I am at a loss for words that would adequately describe the grandeur and the spectacle. It was more of a physical experience. Whether the wind was whipping up the sea, throwing it on the rocks in showers of frothy spray, or it was one of those rare days of dead calm and the surface was merely heaving gently as if a giant was breathing slowly under the surface, it was always stunning. And depending on which way we were looking, the backdrop would be either the wall of the Lofoten Islands' mountains with their jagged peaks or those of the Norwegian mainland across the Vestfjord, more than 15 km away but still immense and unyielding and usually topped with snow.

This long distance played a role in an incredible encounter that a Swedish diver on the boat had with a couple of his diving buddies. For Göran Ehlmé the adventures in Henningsvær and on *Old-Bi* would shape his future, as they did for so many of us. Like Lars Øivind and most of the other Scandinavians, Göran was tall and blond, but in contrast to Lars Øivind, Göran was immaculately crew cut, clean-shaven, and well organized. One evening he showed us pictures from dives in the incredible kelp forests along the Norwegian coasts and it made me understand that a profound love and understanding of the life below the surface was not reserved to biologists and scientists. I can hear that this may sound arrogant, but I was not used to people sharing my love of everything wild and it was a big part of my infatuation with the life on board *Old-Bi* that I had found so many like-minded people.

One afternoon Göran and his diving buddies didn't come home as usual when it began to get dark. This was slightly worrying, and after an hour we started to get nervous. Vestfjord is a huge open sea and not a place to get lost or have engine problems, especially not at night. Luckily it was a very calm evening, so we were not afraid that they had run into bad weather, but there were many other things that could have gone wrong. We couldn't reach them on the VHF radio. We figured that they were either too far away or had run out of battery. Neither was a good situation. Every time there was a sound outside someone hurried from the dinner table to see if they had returned. Finally, we heard the unmistakable noise of an outboard engine and all rushed on deck to meet them. They were laughing and chatting, talking at the same time as each other, over the moon with excitement.

Once we were all back below deck, Göran told us how they had encountered two big male killer whales on the western side of the Vestfjord, not far from the harbor. They were on the move but swimming slowly, he said, and they decided to follow them to see what they were up to. Suddenly, without warning, the whales picked up speed. Mimicking their movements with his hands, Göran described how the whales started to swim so fast that they were porpoising, their enormous bodies coming completely out of the water and making gigantic splashes when they reentered it. Untiring and determined, the killer whales were making a beeline for some unknown destination, not changing direction and not slowing down. In the little dinghy, Göran and his companions had difficulty keeping up with them, despite running the outboard engine at full throttle and making 50–60 kph.

"We had no idea why they were going so fast," Göran said, "and we started to wonder if we should give up and return." But all the way across the huge expanse of the open sea in the Vestfjord, the water was dead calm, luring them on. In the light from the setting sun, the sea looked like a piece of heavy silk that was being gently moved by invisible hands.

Almost across the Vestfjord, just off the islands fringing the other side, the two killer whales finally slowed down, and Göran and the other divers realized what had made them set off on their hour-long journey across the fjord.

"There was a large group of killer whales already there," Göran said, "and they were feeding." He looked up and smiled almost triumphantly. "So I decided to go in."

"Are you crazy?" someone said. "Weren't you scared?"

Göran smiled even wider and replied that no, he wasn't. The animals were so calm and so preoccupied with feeding that it felt safe to go in the water. There was very little light once he was under the surface, but he could still make out the contours of the killer whales as they were moving around what was left of a huge school of herring. The feast was almost over, with many dead herring lying on the surface and lots of fish scales in the water. He stayed with the whales until it was completely dark and the first bands of northern lights started to undulate in the sky, coloring it green and yellow. Surrounded by the killer whales, he had looked up and seen the millions of small silvery fish scales glimmering like stars against the illuminated sky.

As a dutiful cook, I didn't go out on the water as often as the rest of the team. So while everybody talked excitedly of their encounters around the dinner table, I lamented that I still had only seen them once—on that first day. It wasn't until Morten came back early one day because they had forgotten something that I was invited again to join. A small group of killer whales had been spotted very close to the harbor, and he was sure they would still be there if we moved quickly.

I had learned my lesson on that first Zodiac trip, so this time I borrowed one of the many waterproof survival suits before I climbed into the small rubber boat tied up to the side of *Old-Bi*. The boat must have been leaking air because the pontoons were very squishy, and we had to go frustratingly slow. But for a small trip out of the harbor and around the first big headland, it would have to do. When we left the protected inner part of the harbor and turned around the corner to face the open sea, the wind hit us strongly. I knew that Vågakallen, the nearest mountain, was towering over us, but it was hidden in low hanging clouds, and a cold rain beat onto our faces as we approached the area where the whales had been spotted. As we entered the bay below the mountain, we came into the lee and the sea was calmer. But there were no whales to be seen.

We stayed in the bay for a while, looking intently in all directions, but there was no sign of the whales—just the sea swelling with every wave that rolled in and a few seagulls circling above us. Morten shut off the outboard engine and started to rig a hydrophone. It looked like a little round black can and was attached to a long and heavy cable. He connected it to an amplifier and gave me the headphones to listen. Then he slowly lowered the hydrophone into the water. Through the headphones, I could clearly hear the splash and gurgling from the hydrophone as it sunk, and then the quietness of the big sea, with a low thrumming in the background, which I would later learn was the sound of boat traffic in the distance. But through the muffled noises of engines and water, I also heard the most incredible sounds, eerie and melodious at the same time. Like a tropical bird singing a mournful song or people whistling from far away across a deep valley.

Completely amazed, I realized that I was listening to killer whales—the sounds echoed and reverberated in the deep sea. And even if we couldn't see them, they were definitely there, and they were making these astonishing sounds. I looked around to see if I could find them, but the sea looked as empty and lifeless as before. I returned my attention to the headphones and the sounds while the inflatable boat rocked quietly in the waves. Somewhere, in the vast ocean below me, in the great darkness under the leaden surface of the sea, animals were calling and responding to each other. I wondered how far away they could hear each other and what they were saying. Was it the calls of other killer whales that had led Göran's two big males across the fjord? Were they trying to find each other in the depths of the sea? Morten and I took turns listening, but after a while the sounds became weaker and weaker, and after about 20 minutes we couldn't hear them anymore. Reluctantly we started the outboard and puttered back to the *Old-Bi* while darkness descended on us as well as on the creatures below.

Going out in bad weather only to find that the whales that had been spotted from land had disappeared happened all the time. So, with winter's arrival, we locked up the boat and went home, agreeing that we had been exceptionally unlucky with the wind and waves. The following autumn, I once again happily forfeited my university courses in exchange for work in the galley. That season as well, the Center for Studies of Whales and Dolphins had invited the well-known killer whale scientist Dr. John Ford and his wife, Beverly, from Canada to Henningsvær to assist us in our efforts.

We quickly learned that the conditions during our first season had not been that unusual after all, and the same pattern repeated itself.

Countless times we ventured out because we thought the wind was dying down, only to hurry back to port a little while later when it had picked up instead. We were embarrassed that Dr. Ford had traveled so far to get thrown about in the wind and waves or to huddle below decks with us. But he was patient and assured us that we weren't doing anything wrong and that we should persevere. And perhaps also look for whales in other places.

After two seasons, we gave up working in the archipelago around Henningsvær in Lofoten. Whale sightings were not consistent enough to make whale-watching a realistic business. Mic and the biologists on board decided that we should look for whales in the summertime and go even further north, off the island of Andøya. Someone had told us that killer whales were sometimes seen up there in the summer, when the midnight sun made the days eternal and the temperatures tolerable.

I never saw the whales that late afternoon with Morten under the dark skies in the Lofoten Islands, but hearing them shifted something in me. I was hooked. The direction that I had been looking to take with my studies, and even with my life, began to take shape. I was fascinated by how unlike us killer whales are and, at the same time, how oddly familiar. They live a life that is so completely different from ours, most of the time submerged in the coldness and darkness of the sea, an environment utterly foreign to us. But in other ways they are not so alien: they live in family groups, they stay in close physical contact with their kin, and they communicate with each other. They are social beings just like us. I was deeply moved when I realized that human beings share such fundamental characteristics—like our ability to communicate and care for each other—with animals that are as foreign as killer whales.

My own fascination with killer whales, too, left me unprepared for the strong emotions of fear and hatred they elicit in many people.

During the unsuccessful years in the Lofoten Islands, when we were still trying to locate a good spot to study the animals, I sometimes met local people, especially fishermen, who declared them to be robbers and thieves, pests, or menaces that should simply be wiped out. And these weren't just verbal threats—on a few occasions, I witnessed a fisherman take a rifle and the law into his own hands and shoot at killer whales from his fishing boat.

The mixed emotions that killer whales are met with are not reserved to this species; they are common around all kinds of apex predators, including bears, wolves, lions, crocodiles, and hyenas. When I am not in Norway, I live in Denmark, and I experienced this potent mixture of fear and admiration close at hand recently with wolves. Although Denmark is a neighboring country to Norway, it doesn't share many, or any actually, of the natural wonders that you find in Norway. There is no wilderness, there are no mountains, no deep fjords, and no glaciers, and the little wildlife that remains is cornered in small enclaves of woodland or moorland pinched between the intensively farmed landscape and the highways paved with asphalt. So it was bombshell news when a wolf was sighted at a lakeside in Jutland in western Denmark. It had sauntered over the border from northern Germany, where wolves thrive in the great expanses of heathland and rough grazing between Hamburg and Hannover. It appeared precisely two hundred years after the last wild wolf had been shot in Denmark in 1812, and its sudden presence raised the hackles of those who thought wolves still had no place in a small and densely populated country like Denmark. Though it also sparked the enthusiasm of others who thought that wolves were a wonderful addition to the country's meager biodiversity.

The rogue wolf, unfortunately, died of natural causes, but shortly thereafter a few more wolves were sighted in Denmark, and they just kept coming in from Germany. A few more years passed and Denmark's national nature agency reported that there was a breeding pair in the western rural part of the country. Eight wolf puppies were born and thrived well, but a hot-tempered debate arose in national and social media between the nature lovers, the farmers, and other locals afraid to let their children outside to play. The occasional attacks and killing of farm animals, especially sheep, added fire to the arguments.

Then the tone of the debate suddenly went from bad to worse. One spring morning, a couple of wildlife enthusiasts were filming one young female wolf from a camouflaged hide in some woodland bordering a more open area. The slender, beautiful animal trotted calmly across the field in broad daylight. In the background, a red tractor is plowing, and a chaffinch is singing peacefully from a nearby bush. A car approaches from the other side of the field and comes to a stop. The driver then pulls out a rifle and coolly shoots the wolf, which falls to the ground, her legs twitching a few times. She bites her own side where the bullet entered, in obvious anguish, before dying. Then the man in the car drives away. In the film, you hear one of the observers cry out, "He f— shot her!" And before long the video was widely shared in the media. That and the subsequent trial of the shooter added even more tension to the debate, which turned bitter and toxic.

In many areas where wolves live close to people, such conflicts are common. In places where killer whales live close to people, they are—or have been—common too. Like wolves, killer whales are found worldwide, and although one species lives on land and the other in the sea, both are highly adaptable and found in many different habitats. As apex predators, killer whales and wolves are at the top of their respective food chains with no natural enemies. They hunt in packs, and they are smart—though some say intelligent, and some say cunning. They exploit new territories and new possibilities quickly and they have little to fear.

Just as the wolves are not afraid to take from the silver platter of a farm's unguarded sheep or cattle, so killer whales also interfere with human business. Whalers see them as insatiable murderers, ironically criticizing them for exactly the same actions as they carry out themselves, and fishermen complain both that killer whales destroy their nets and gear and that they deplete the ocean of fish. We frequently heard these arguments from locals in the pub in the early years on the Lofoten Islands.

But in other places, killer whales have attained a strikingly different status. In the Pacific Northwest of the United States and Canada, they are revered as icons of the region. In 2007 the governor of Washington State declared June to be the official "Orca Awareness Month" in order to "focus awareness on the plight of the fragile Southern Resident community of orcas, to honor their presence in our waters and to speed up efforts to recover the population." This striking declaration has been endorsed and reconfirmed in each subsequent year, as well as by the governors of neighboring Oregon and British Columbia, who later joined the initiative. Their public statements and the associated festivals and celebrations demonstrated an unambiguous devotion to the local whales.

"We are blessed to have this urban community of orcas in our midst," said one of the proclamations. While another admitted, "We have only begun to learn about the intelligence and social capabilities of orcas." A third emphasized the cultural importance of killer whales to the indigenous people in the area.

These political declarations reflect the role killer whales play in many aspects of life in the region, including the connection local people feel to them as well as the impact they have on tourism, the economy, and the ecosystem. They also reflect the concern there is for the local whale populations, which have been declining and are now reaching critically low numbers. Each year local residents and visitors celebrate the return of the whales to their area with festivals and songs, and if a killer whale dies, they mourn the death as if it was a family member.

The contrast between this modern-day view on killer whales in Canada and the United States and the way they were seen in the Lofoten Islands in the latter part of the last century is stark, but it was only a short time ago they were regarded as a pest in the Pacific Northwest as well. As I dove deeper into the world of killer whales, I began to understand that these extreme attitudes are part of the fate that follows killer whales everywhere, just as it does wolves and many other apex predators.

This book grew out of my fascination with killer whales and my desire to better understand the role they play in the lives and imagination of people around the world. How are they perceived as a threat and a competitor in some places while revered and admired in others? What determines this difference in attitude and understanding? How did their reputation change from bloodthirsty killers and savage monsters that should be eradicated to the complete opposite over the last few decades in the Pacific Northwest?

My journey to understand this started in that small leaking inflatable in the choppy waters off the Lofoten Islands. It has taken me across continents and oceans and into the history and lives of people living by the sea and from the sea. Although I have studied killer whales in the wild and I am a trained scientist, this is not a scholarly work. It is a patchwork of stories I have collected over my years on the ocean about our relationship with the biggest predator on Earth. Stitched together they form not a precise and finely drawn pattern but a mosaic of the different hues and tones that history and culture are made of.

Killer whales are unconcerned with our attitudes. They don't need our love or our hatred. How we understand and interact with a big predator like the killer whale is instead a reflection of ourselves and how we want to live with the complexity of other animals around us.

CHAPTER 1

Bloody Beasts

Killer whales or orcas: what's in a name? And does calling them one thing and not the other put them—and yourself—into a certain category? To many, especially people involved in killer whale conservation, they are and should be *orcas*. Calling them killer whales, they say, labels them as bloodthirsty and cements the perception that they are merciless monsters. The conservationists argue that calling them orcas instead leaves room for a more nuanced picture of the species and its behavior, which doesn't label them as killers. And they are right of course, because killer whale behavior is highly diverse and full of complexity, including caring for their young, being playful and tender, and maintaining lifelong relationships with their family members.

Still, I call them killer whales, and I apologize if this is offensive. I do it partly out of habit (this is what the scientific community I first studied with called them) and partly because their scientific name, *Orcinus orca*, to me is not much better—it means "the demon from the underworld," as Carl Linnaeus was well aware when he christened them.

Nobody knows when killer whales were first identified as a species, but with their striking coloration and formidable size it seems likely that they would have been recognized very early. Apart from rock carvings, which of course also qualifies as a kind of description, we owe the first written description of killer whales to Pliny the Elder, who lived in the first century AD.

Pliny was a diligent and hardworking officer of the Roman Empire, who in his spare time sought to collect and present all the knowledge in the world into his magnum opus *Naturalis Historia* (Natural History). It is sometimes called the first encyclopedia as it deals with almost everything in the natural world: astronomy, mathematics, botany, geography, pharmacology—you name it. It is all in there together, along with human activities like mining, sculpture, painting, and agriculture. *Naturalis Historia* is not organized alphabetically like a modern encyclopedia but presents itself as a guided tour, like an ancient precursor to Bill Bryson's enlightening and entertaining *A Short History of Nearly Everything*. *Naturalis Historia* is not short though (neither is Bryson's book, come to think of it). Pliny's comprises no less than 37 volumes.

Killer whales appear in volume nine, "The Natural History of Fishes." Here Pliny dedicates a whole chapter to whales, which at that time were classified as fish. That was not as such a bad mistake as it sounds despite that Pliny and others before him knew very well that whales were mammals that nursed their young from milk glands, breathed air with lungs and not gills, and had forelimbs instead of fins. This is something they had probably learned from studying dead whales and dolphins found on the shore. The fact that they were marine creatures living in the water was the determining factor—they belonged with the fishes. But they were still different enough that Pliny grants them a chapter of their own apart from the other fish.

Pliny tells how big baleen whales sought shelter in secluded bays to give birth to their calves and then adds: "*This fact, however, is known to the orca, an animal which is peculiarly hostile to the balæna, and the form of which cannot be in any way adequately described, but as an enormous mass of flesh armed with teeth.*" Pliny asserts that the desperate whales

> *are well aware that their only resource is to take to flight in the open sea and to range over the whole face of the ocean; while the orcæ, on the other hand, do all in their power to meet them in their flight, throw themselves in their way, and kill them either cooped up in a narrow passage, or else drive them on a shoal, or dash them to pieces against the rocks.*[1]

Pliny doesn't describe in detail what killer whales looked like and it is possible that he had never seen them himself but was relying on descriptions from seafarers and others who had come across them. Pliny's portrayal of killer whales as monstrous was an image that would cling to them for centuries.

The killer whale's behavior when hunting larger prey was also noted in another ancient book. Somewhere between the years 1250 and 1260, the Norwegian King Håkon Håkonsson had a book made with instructions to his sons about his kingdom and how to rule it, a kind of fatherly introduction to the ins and outs of being a king. The book is called *The King's Mirror* (Kongespeilet) and includes in-depth information about the furthest reaches of his kingdom, including Iceland and Greenland—handy knowledge for Håkon's descendants to enforce their sovereignty in such an extensive realm. The animals in the sea get special attention, and a varied and detailed knowledge of many species of whale is included in the book. For a seafaring people like the Norwegians, this was vital information. Killer whales are featured for their gluttonous appetite:

> *There is another kind of whale called the grampus, which grow no longer than twelve ells and have teeth in proportion to their size, very much as dogs have. They are also ravenous for other whales just as dogs are for other beasts. They gather in flocks and attack large whales, and, when a large one is caught alone, they worry and bite it til it succumbs.*[2]

Like Pliny, the author of *The King's Mirror* considered all whales, including killer whales, as fish and they continued to be classified as fishes for quite a while. When the renowned taxonomist and natural historian Linnaeus decided to name all living organisms in the world, he listed the whales along with the fishes too—at least in his early treatise from 1746. He was also the first to give killer whales a scientific name. He called them *Orcinus orca*, keeping the name *orca*, which had been around at least since the time of Pliny, and adding *Orcinus*, Latin for "belonging to the underworld." The meaning of the word *orca* is a bit obscure, but it may be derived from *orcus*, meaning "underworld." Others think it comes from the Latin word *orca* for "barrel" or "cask," referring to the shape of a whale's body.

When Linnaeus published the tenth edition of *Systema Naturae*, he made a quick decision and moved all the whales and dolphins, including the killer whales, to be grouped with the other mammals and not with the fishes. He placed the killer whales in the family of dolphins (Delphinus) together with pilot whales, bottlenose dolphins, and common dolphins. Modern taxonomists, who have studied both the anatomy and genetics of this diverse family, agree with Linnaeus on this; killer whales are essentially very large dolphins.

Local people in areas with killer whales, as well as sailors and whalers, knew these animals well, of course, and had a wealth of names for them, like *blackfish* in Canada, *spekkhogger* in Norway, and *ardlursak* in Greenland, but natural historians continued to struggle with how to identify and characterize killer whales despite Linnaeus's efforts. The big difference in size between the dorsal fins of the males and females, for instance, was a source of confusion. Were there two different species or a single one? And the many different reports on the coloration also perplexed the natural historians who were so eager to assign every species to a designated space in the order of life. Sometimes killer whales were black and white, while at other times they were described as being black and yellow or with violet coloration on their flanks. The confusion was undoubtedly aggravated by descriptions being made of dead animals, as the color of a whale changes rapidly after death as decomposition sets in.

Much of this uncertainty waned when the Danish zoologist Daniel Eschricht performed an autopsy on a dead killer whale in 1861. After the dissection he wrote a detailed description that clarified most of the ambiguities. But his report became famous for a very different reason and helped to sustain the bloodthirsty reputation of killer whales. He claimed that in the stomach of the dead whale he found the remains of no less than 13 seals and 14 porpoises. His observations are still cited all over the world and continue to raise eyebrows. Could his report really be true, or did he wildly exaggerate?

After the autopsy, the skeleton of the whale was shipped to the Natural History Museum in Copenhagen, where it still resides. I asked the museum's conservator, Abdi Hedayat, if I could come see it and he generously welcomed my visit. We met in the foyer of the museum, packed with excited preschoolers waiting to go inside. We weren't going into the public part of the museum, though, but into the well-hidden collection rooms.

Abdi is a whale specialist and has led a number of dissections of cetaceans—the taxonomic group of mammals to which all species of whales, dolphins, and porpoises belong—some much bigger than killer whales, like bowhead whales and fin whales, and some smaller ones, like dolphins and minke whales. He gained nationwide fame when, in a live television broadcast, he punctured the swollen stomach of a stranded and very dead sperm whale, only to have the rotting innards explode, drenching him in a cascade of incredibly smelly fluids. His laughing face, splattered with red gore, made him instantly popular and respected.

"Cutting up a whale today is not much different from when Eschricht did it," Abdi tells me, as he unlocks the first of several doors between the foyer and the collections rooms. "It's still a question of using a sharp knife and coping as best you can with the foul stench."

As we approach the collection room for big mammals, we're met with a distinctive smell here, too, though from a different source. It's the unmistakable odor of whale bones. It's not unpleasant like the overpowering stench of a rotting whale, but it is pungent and pervasive, oily and fermented, with a hint of something fishy. Abdi punches a code into a small keypad and opens the door to a huge room. As we step inside, the smell intensifies.

"If you work in here for a whole day, you'll have to wash all your clothes afterwards," Abdi explains, as we walk along the many shelves and old wooden display cases. "It lingers in your clothes and not everybody likes it . . . my wife certainly doesn't," he laughs. Personally, I like it. I imagine it is what an old warehouse in the days of whaling would smell like, full of tar, coiled up ropes, and barrels of whale oil.

In the collection room, there are bones and skulls and heaps of ribs and vertebrae everywhere. Some of the skeletons are mounted on old handmade wrought-iron stands. A handwritten label with elegant letters in black ink identifies one of them as a pilot whale from Vaagø in the Faroe Islands, collected in 1844. The jaws are locked in a perpetual grin of worn-down and crooked teeth. Next to the pilot whales, on a low wooden cabinet, rests the complete skeleton of a narwhal, with not just the usual one but two enormous, spiraled tusks. The label on this specimen informs me that it's from Greenland and was secured by the museum in 1847.

Most of the skeletons are not mounted. The biggest ones lie on the floor, the smaller ones on wooden shelves or in Styrofoam boxes piled up in stacks. The huge skull of a sperm whale is almost blocking our way. The bone of the skull is grey and porous, weathered from lying outside for a long time before it made it to the museum. I touch it gently. It's coarse and grainy, like driftwood. Similar to wood, bone seemingly transform into a different material if left outside for long enough.

Abdi leads me to the shelves with the killer whale specimens. They're all lined up with the skulls pointing the same way and the vertebrae in a long line behind each. At first glance they look like a parade of crocodile skulls with their gigantic conical teeth.

"Here is Eschricht's killer whale," Abdi says, as he stops by a very large specimen. "It was the museum's first killer whale specimen, so it has collection number one."

He explains that each specimen in the collection room has a number, and the number is inscribed on every single bone from the same animal, ensuring that, should the different parts become separated, they can always identify to which animal a bone belongs in the museum's catalog. This is crucial information when doing research, along with the catalog's information detailing where the animal was found and who collected it, plus any other relevant information. Without this, a bone—or any other part of an animal in a collection—is worthless for scientific purposes.

Eschricht wrote down every detail about the killer whale and presented it at a lecture. He was a professor of physiology and comparative anatomy at the University of Copenhagen, and like many of his contemporaries, he also collected natural history specimens. He had a particular, if slightly unpractical, soft spot for whales and whale skeletons. For decades he collected everything he could get hold of, and by virtue of good connections between Denmark and Greenland, which at that time was a Danish colony, he managed to stockpile the skeletons of many different species of whales—most of them big, bulky, and smelly.

From time to time, usually following a complaint from his wife, he promised himself that he would stop collecting. He seemed unable to stick to that repeated resolution, though, and the skeletons and skulls and the preserving jars full of parasites and organs and fetuses of all kinds of animals continued to accumulate in his home. Eventually his wife had had enough, and the skeletons and specimen jars were donated to a newly established private museum. From here they were bought by Copenhagen University's Zoological Museum in 1841.

Eschricht didn't just collect the whale specimens, either. He also wrote scholarly dissertations about them, on porpoises, beaked whales, bowhead whales, and killer whales. It is in recognition of his early and profound contribution to whale science that in 1860 the British scientist John Edward Gray named the grey whale *Eschrichtius robustus* after him.

Eschricht himself carefully documented the story of Specimen One, the killer whale skeleton, in the report he wrote afterward.[3] It started on July 21, 1861, around noon, when Eschricht received an unusual telegram from Count Ernst Benzon, who resided at the Benzon Castle in Jutland, many hours travel from Copenhagen. During a sailing trip in the waters near his estate, the Count had come across the drifting carcass of a dead whale and had towed it to land. If Eschricht would come right away and take care of the business, Benzon was willing to donate the whale carcass to the museum. He only had one stipulation: he would retain any valuable blubber or oil from the whale for himself.

Eschricht didn't hesitate. He jumped into action immediately and by five o'clock on the same afternoon he was boarding the steamship from Copenhagen. To assist with the unpleasant job ahead, he brought with him an early holder of Abdi Hedayat's post, the preparator Mr. Iversen from the Natural History Museum at the university. When they arrived in the harbor at Grenå the next morning, the Count's men met them with a horse carriage and took them directly to the beach where the dead animal lay.

As they approached the site, they could smell the rotting whale even before they saw it. Eschricht feared that the carcass would be so decomposed that a scientific investigation would be impossible. But as it came into view, his hopes rose, despite the overwhelming stench. From the black and white coloration of the dead animal, he knew immediately that it was a killer whale, a species he had never seen before but which he recognized from illustrations, and it looked intact. Reading Eschricht's report, it's almost possible to hear his heart skip a beat in anticipation. One more skeleton for the collection, and a new species too!

Eschricht later learned that a killer whale had stranded on a beach in Denmark at least once before, but that one had never made it to a museum. An oil painting of that unfortunate whale hung on the walls of the grand house on an estate neighboring Benzon's. Apparently, it was still alive when it stranded, but the accompanying text on the painting left no doubt about its fate: "*The year 1679, on December 27th, this fish landed at our ferry jetty in the fjord. Here it was shot by the Right Honorable Christen Seefelt of Steenalt, first with a musket bullet in the head and thereafter with nine big bullets in the stomach until it died and could be salvaged.*" Eschricht politely asked for a copy of the oil painting and was delighted when he later received a beautiful watercolor reproduction. Although it was unmistakably a killer whale, the scientist in Eschricht couldn't help but note that the proportions were wrong and that it was not quite anatomically correct.

This was not a mistake Eschricht would make himself. He evaluated the dead whale and, by the very visible massive penis, immediately recognized that it was a male. (In a dead whale, the penis often protrudes, making sexing easy even for beginners.) The dorsal fin was tall, 6¼ feet or almost 2 meters. Eschricht also noticed that the pectoral fins were very large. Assisted by Iversen, he performed a meticulous series of measurements of the animal: its length, the length and width of its pectoral fins, the distance from the mouth to the eye and from the eye to the ear, the distance from the tip of the snout to the blowhole and back to the dorsal fin, and not forgetting, of course, the length of the penis and its circumference.

I am emphasizing Eschricht's habit of taking precise measurements because his observation of the heaps of dead seals and porpoises inside this dead killer whale is quite astounding and has often been questioned by later scientists because they thought he must have been exaggerating. Yet, reading his descriptions leaves no doubt about his careful observations or his dedication to making an honest report of his findings.

When Eschricht and Iversen were done with all the measurements, Eschricht turned his attention to the mouth of the whale in which a dead seal was stuck, clutched firmly in its jaws. Eschricht had already been told about this by the people who had picked him up when he arrived; they suspected that the killer whale had somehow suffocated on the seal. Eschricht struggled to pull the seal loose from the vicious-looking teeth but eventually managed to extract it. He found that it was only the skin of a seal, turned inside out, with the claws sticking out from a hole in the skin. Upon further inspection, the skull of the seal was also found on the inside of the skin.

Now the men on the beach were ready to start cutting away the blubber and flesh of the whale. They cut the blubber into pieces and brought it to waiting barrels. The meat they discarded, pulling it with hooks and ropes back to the shore where they tossed it into the sea. A handful of men, presumably Benzon's farmhands, assisted Iversen with the arduous and dirty work. Meanwhile, Eschricht busied himself by studying the head and genitals of the dead whale—or it is also possible that he was not too keen on being involved with the actual dismembering of the stinking carcass.

As the day neared its end, the men were finally ready to cut open the stomach and investigate the insides of the whale. With a long slash they cut the belly open and as the internal organs spilled out onto the beach they became immediately aware of the size of the stomach, which seemed enormous. Inside the first chamber of the stomach they found an intertwined mass of dead animals. Eschricht couldn't believe his eyes. In his report he wrote: "*Even if I was prepared to find something extraordinary inside the stomach, I was still utterly amazed to immediately find five or six seals, some small, some larger, entangled in each other. To count them it was necessary to get them out and separate them.*"

But Eschricht was in for more surprises. As they pulled out the dead seals, they discovered that, underneath them, even more seal carcasses were hiding. And below those were the semi-decomposed carcasses of just as many porpoises. Eschricht approached the entwined mass of bodies methodically.

"*In order to determine the exact number of dead animals,*" he wrote, "*it became necessary to place them in rows on the beach.*" Because the porpoises were so rotten, he counted the skulls to make sure that he was not overestimating the number of dead animals. With all the dead seals and all the porpoise skulls lined up, he was able to make a count: "*In the stomach there were 13 porpoises as well as 13 seals; a 14th seal which was very small but intact had passed from the first compartment of the stomach to the second.*"

He speculated that the skin he found stuck in the jaws of the killer whale could belong to one of the carcasses and therefore did not include it in his total count of 27 seals and porpoises. All the dead seals in the stomach of the whale were without fur. Later observations have confirmed that sometimes, if not most times, killer whales remove the skin of a seal before eating it.[4] Perhaps the vigorous "beating up" that has sometimes been filmed, when a killer whale is throwing a hapless seal into the air, repeatedly bashing it with its tail fluke, is not so much a cruel act of play as it has often been suggested, but is done to loosen the skin so it can be removed before the seal is ingested.

As darkness enveloped the beach, Eschricht hurried to get the intestines out and measure them. They reached an impressive length of 177 feet and 8 inches—or a bit more than 54 meters. Intestines measured, Eschricht called it a day and he and Iversen left for the hospitality of Count Benzon, who invited them to recover from the hard work for a few days at his estate. Before leaving, Eschricht made sure that some of the farmhands stayed by the remains of the whale during the night so that nobody could run off with his prized skeleton.

As agreed, Benzon got all the blubber. It yielded 210 pots of oil, or around 200 liters, worth a small fortune of 87 rigsdaler, or probably more than four months' wages for one of the farmhands assisting in the work on beach.[5] But Eschricht didn't care about the oil, he wanted the skeleton for the museum, and he started to worry that it might be too stinky to be allowed on the steamship back to Copenhagen. So the following day he cut short the recovery time allowed to poor Mr. Iversen and instructed him to continue cleaning the bones of the whale, while he himself recuperated in the pleasant company of the Count.

Eschricht's thoroughness and Iversen's hard work paid off. Eschricht doesn't explain how they managed to bring the collection of bones on board the steamer going back to Copenhagen, but manage they did, and to this day it still resides in the Natural History Museum's collection, where generation after generation of conservators have taken good care of it.

The skull of killer whale number 1 is impressive; it must be one of the biggest in the collection. The teeth in the front are low and rounded, a testament to their daily use in the long life of the whale. In the back, they are more pointed still. One tooth is missing, but it is hard to know if this happened before the whale died or after. I lean in closer to the skull. The smell of whale oil still lingers in the bones, but it is faint and mixes with all the other smells in the collection room.

"Eschricht's observations must have made headlines in the scientific community then," Abdi tells me. "Once when I was reading a bedtime story to my kids, I came upon the many dead seals and porpoises Eschricht found in the whale's stomach in a very unexpected place." Abdi explains that he was reading Jules Verne's story *Twenty Thousand Leagues Under the Sea* when he found this passage:

> *Schools of elegant, playful dolphin swam alongside for entire days. They went in groups of five or six, hunting in packs like wolves over the countryside; moreover, they're just as voracious as dogfish, if I can believe a certain Copenhagen professor who says that from one dolphin's stomach, he removed thirteen porpoises and fifteen seals. True, it was a killer whale, belonging to the biggest known species, whose length sometimes exceeds twenty-four feet.*[6]

Clearly Verne, who was very well educated, must have read about Eschricht's discovery. *Twenty Thousand Leagues Under the Sea* was published in 1869, seven years after Eschricht's report. When Abdi noticed this similarity, he looked into who else had commented on it but found nothing. Apparently, nobody had noticed the connection between the two before Abdi's observation, which he later published in the museum's natural history magazine.[7]

Undoubtedly Eschricht's astonishing account of the 27 seals and porpoises in the stomach of the killer whale cemented the perception of killer whales as ferocious, gluttonous monsters and thereby influenced our view of the species and blocked our understanding of it for a very long time. Although Eschricht's observations have sometimes been questioned, his precision and obsession with details leaves no doubt in my mind that, at least in this case, the killer whale's reputation for gluttony was not wholly unwarranted.

CHAPTER 2

Sea Change

When the *Old-Bi* left idyllic but windblown Henningsvær to search for killer whales further north, I decided to stay aboard. We had closed everything down and left the boat in Henningsvær for the winter, but with the same sure instinct as migrating birds we all headed north and congregated again when spring arrived. By then I had added a few other dishes to my repertoire in addition to marinated herring and was competently making bread, cinnamon rolls (always popular), and a variety of pasta dishes in the small galley—competently enough that I was welcome to continue as a crew member, anyway. It was an easy choice for me to continue sailing and postpone my studies (again).

We arrived at Andenes on the northernmost point of Andøya at midsummer, after two days of sailing in glorious weather. The contrast to Henningsvær in the fall was stark. On the one hand there was the midnight sun and blue skies instead of wind and rain. But on the other,

Andenes was also a much bigger and less picturesque town than Henningsvær, with its narrow streets and a harbor full of wooden fishing boats and old warehouses on stilts. In addition to hosting a big military airbase, Andenes had a large harbor surrounded by fish factories and industrial buildings, and it was laid out on a grid with long, straight streets and square modern houses. But it made up for its lack of charm by being surrounded by a majestic landscape with massive mountains, a ragged coastline, and long sandy beaches.

At this time, hardly anything was known about killer whales in Norwegian waters. Even basic things like how many there were or what they ate (other than herring) were unknown. The few published reports available presented anatomical measurements based on whaling data but no information on behavior or social biology. Contemporary studies of killer whales on the other side of the Atlantic, in Canada and the United States, had shown those killer whales to have intricate social systems and complex vocalizations by which they communicate underwater. These studies of killer whales and the increase in knowledge about them had already helped transform attitudes to them in the Pacific Northwest. The understanding that the whales were intelligent and lived in stable, vibrant social groups contributed greatly to replacing fear and loathing with interest and fascination.

Several of us on board *Old-Bi* were budding scientists, and to us these studies were hugely inspiring. How strong are the social bonds between killer whales in Norway, we wondered? Do they also live in stable family groups, or is theirs a more fluid society? Are they all herring eaters, and do they communicate in the same way as in the Pacific Northwest? We were eager to follow in the footsteps of the American researchers, to start a similar study in Norway, and to begin to understand the Norwegian killer whales better. Our own encounters with fishermen had underscored the need for a change of attitude in Norway, something that we were eager to contribute to as well.

From Andenes, the North Atlantic Ocean stretched westward all the way to Greenland, offering no protection from the elements. Rainstorms and rough seas were common here even in summertime. However, when we arrived it was calm and sunny for once. In the gorgeous weather, we stayed in harbor just long enough to get some fresh provisions and then we headed out to sea again in *Old-Bi*, to take advantage of the good conditions for as long as they might last. On that first trip out, we didn't see any killer whales, but to our delight and surprise, we found sperm whales instead. We came upon several large adult males, spaced half a mile or a mile apart. If we had known better, we wouldn't have been so surprised; sperm whales are a deepwater species and males are known to migrate from the warm tropical and subtropical seas where the females and calves live, to more northerly latitudes to feed.

We soon learned that the sperm whales were regularly found only six to seven miles from Andenes, as the continental shelf plunged steeply at that point into a deepwater trench. During the following weeks, we found the sperm whales there every day we went out. They were large, from 12 to 16 meters long, and would lie still at the surface, exhaling steamy plumes of spray from their blowholes—which is known simply as "blowing." They blew regularly for almost 10 minutes after every 45-minute dive. While resting like this, they could be approached carefully, and at closer hold we could see their immense size, the peculiar blowhole at the tip of the snout (and not on top of the head as in all other whales), and the heavily wrinkled skin that is characteristic of the species. When they were done refilling their oxygen supplies, they lifted their flukes (tail fins) and gracefully dove back down into the abyss.

Occasionally we did encounter killer whales, too, but those occurrences were more unpredictable than of the sperm whales, and our idea of establishing whale-watching in Norway took a new direction. Instead of basing it on killer whales, we put the much more reliable sperm whales at the top of the list, and soon we started trying to convince local people of the value of the idea. One step was to invite the local authority's director of economic development, Bjørnar Sellevold, to the *Old-Bi* to talk about the potential for whale-watching in Andenes.

Bjørnar arrived with a bagful of school buns—Norwegian pastries filled with custard and dipped in sugar—from the local bakery, jumped on board, and greeted us enthusiastically. Full of questions, he sat down at the small wooden table in the cabin and wanted to know where we came from, why we had left Henningsvær, and what we had discovered in Andenes.

"Is there anything special here?" he asked.

When we told him about the sperm whales, his jaws dropped. He had no idea that there were big whales to be found so close to his hometown, but he immediately realized the potential for developing tourism and combining whale-watching and research.

He found the sperm whales' name hilarious (it is the same in Norwegian), and when we explained that the name had been given long ago by whalers who thought that the many tons of valuable milky oil in the sperm whale's head was the animal's sperm, he slammed his hands on the table and roared with laughter. The next day we read the headline in the local paper: *Whale sex in Andenes*. Bjørnar had made sure that the joke was lost on no one and had gone straight to the editor of the local paper.

The same summer we also invited American whale-watching captain Al Alvelar to visit us in Andenes. Alvelar had become a millionaire taking people to see whales off Cape Cod in the northeastern corner of the United States. Luckily, the weather cooperated during his visit, too, and both local and national journalists spent a beautiful day on the *Old-Bi* watching sperm whales and killer whales.

Alvelar's testimony about whale-watching, and especially his financial success with it, made headlines in the newspapers and prompted an immediate reaction from an unexpected place. Ragnvald Dahl, a young skipper and third generation whaler from the Lofoten Islands, promptly retorted that if anyone was going to get rich on whale-watching in Norway, it should be a whaler. He suggested himself. And this is how whale-watching in Norway started out, as an unlikely alliance between a hot-tempered whaler and a handful of biologists, students, and artists.

It was a volatile combination, and sometimes tourists with anti-whaling sentiments provoked flare-ups. Once Ragnvald declared t-shirts with "Save the Whales" messages unwelcome on his ship, and on another occasion—to the horror of the whale-watchers on board—he used his VHF radio to guide a nearby whaler to a minke whale that had been spotted from the whale-watching boat.

More often Ragnvald was in good spirits and his experience in spotting whales was indisputable. In the end, though, he found it too difficult to reconcile his own attitude and values with being a whale-watching skipper, and he left Andenes again—or perhaps he realized that it would take longer to become a millionaire than he had counted on. Either way, some locals had embraced the idea of whale-watching by then and formed a limited company, taking over the whale-watching business in Andenes.

The sperm whales were as regular as clockwork. The deepwater canyon outside Andenes is a supreme habitat for them, and it was possible to spot them every day through the summer if the weather was not too windy. More and more tourists started to find their way to this remote place, and with funding from the World Wildlife Fund, a small Whale Center was founded. I started working on the whale-watching boats as a field guide. In the first year, the number of visitors could only be counted in the hundreds, but that was enough to make it worthwhile to run trips out to sea every other day or so.

It took time for the local tourist industry to understand the draw of the whales. One day I overheard the local high school student who had a summer job in the tourist office explaining to a young Dutch couple what the island had to offer. They had arrived on bikes and were looking for something to do.

"Something outside? Something with nature?" they wondered.

There wasn't much to do in Andenes they learned from the receptionist, so they left the office with the information that the local sea fishing festival was due to start in two weeks' time. When the door shut behind them, I asked the girl in the reception why she had not suggested that they could go whale-watching that very afternoon; I knew there was a trip scheduled. Reluctantly, she admitted that she thought it was an embarrassing suggestion to make.

"Who would want to do that?" she said, visibly uncomfortable. "It's just whales. It's like saying that they could go watch seagulls or something." I suspect she wasn't the only local who felt that way. But the stream of visitors continued to grow, with or without the support of the tourist office, and gradually they warmed to the idea as well.

When *Old-Bi* left Norway in August for its home port of Gothenburg and other adventures, I decided to stay in Andenes and moved into the Whale Center with a handful of people who either had come with *Old-Bi* like me or had been drawn by the activities and arrived independently. The Whale Center was situated in a wooden house in the harbor, a so-called *rorbu*, constructed at the water's edge for seasonal use. Rorbus were originally constructed for fishermen who came from far away to fish in the rich waters in northern Norway, at a time when fishing boats were open air and therefore unsuitable to live and sleep in.

The rorbu at the Whale Center was yellow with red window frames and had lots of small bedrooms plus a communal kitchen. When we moved in, there were still old fishing nets, empty bottles, a broken TV, and other remnants left by former occupants, but after we had cleaned it up, it lent itself perfectly as accommodations for students and guides.

Biology students and other young people quickly found their way to Andenes. As if by magic, they materialized out of nowhere with their backpacks and volunteered to work in exchange for a place to sleep and a little bit of food.

Once I woke up in the middle of the night with the sun shining brightly through the windows. I gave up sleeping and went downstairs to the kitchen to discover that I was not the only one awake. Three American guys, undergraduate students who worked as guides on the whale-watching trips, had taken over the kitchen table and were manufacturing a hydrophone from scratch, using parts they had bought in an auto repair shop. A soldering iron, tangled cables, and electrical wiring mingled on the table with the leftovers from dinner. Two others were working in a small makeshift darkroom (yes, this was before digital photography) developing identification pictures of killer whales and sperm whales to use in research.

The Whale Center had turned into a vibrant hub for students interested in whale research, and the discussions over the breakfast and dinner tables became the stimulating intellectual environment I had missed at university, greatly expanding my understanding of what was interesting and possible to study. The memory of the underwater sounds of killer whales had stayed with me since the dark afternoon when I first heard them in the waters surrounding Henningsvær, and I had often speculated about their meaning. Like a little restless worm, an idea started to wriggle in my mind: the thought that perhaps I could start a study of killer whale sounds in Norway.

Tiu Similä, a 26-year-old graduate student, was thinking about ideas for research as well. She had also arrived in Andenes on board the *Old-Bi* and stayed when it left. Originally a phytoplankton specialist, she decided to exchange some of the ocean's smallest organisms for some of the biggest and left a promising career as a marine ecologist in her native Finland in exchange for an uncertain life as a whale biologist in

Norway. Unlike all the tall and blonde Scandinavians, she was petite, with chestnut hair and brown eyes—and a strong will. Despite all the unknowns, she was determined to establish a study of killer whales in northern Norway. Our experiences in the Lofoten Islands had made her realize that practically nothing was known about the Norwegian population of killer whales—except that they were there. She wanted to change that.

How to study whales was not something that any of us had learned at university. During the twentieth century, research in biology had taken a swing into the labs. The future for an ambitious biologist was in vials and petri dishes, in controlled experiments and measurements of physiological parameters in cells and organs. This was the kind of biology that many of us had encountered so far in our education. The more traditional methods of observing animals and other organisms in the wild, with roots firmly embedded in the nineteenth century and epitomized by legendary figures like Charles Darwin, Alexander von Humboldt, and Alfred Russel Wallace, were looked upon as a thing of the past. It wasn't real science, it was "just" natural history, and there was little funding and little prestige in being a wildlife biologist.

A few mavericks continued unperturbed down this road anyway. Dr. Jane Goodall's groundbreaking and decade-long study of chimpanzees, starting already in the 1960s, is, of course, well known. What we know of chimpanzees today and how their intricate social life works—with hierarchies, alliances, friendships, fights, and even wars—could never have been understood if it wasn't for Goodall and her later colleagues' long-term observations in the wild. But she wasn't the only one. Not far from Jane Goodall's camp in Gombe, Ian Douglas-Hamilton had been studying elephants for almost as long. He, too, and those who followed in his footsteps, had uncovered a strong and sophisticated social system in elephants.

Elsewhere in Africa and a bit later, Craig Packer went for the lions and Jeanne Altman and Robert Sapolsky investigated the baboons. In the United States, David Mech started studying wolves in the wild while others looked at bears. Even ravens got their own long-term study when Bernd Heinrich made them the focus of his scrutiny. All of them combined long-term studies, involving not days or years but decades of observations, with an effort to recognize and track the behavior of individual animals. As a result, they were able to understand details of social organization and behavior that would have been impossible otherwise. To take this approach with whales, living mostly submerged in the dark sea, seemed like an impossible, even crazy, idea. Yet it happened, and just as it did with the chimpanzees, elephants, lions, baboons and wolves, it revolutionized our view of them.

The first person to immerse themselves in the world of killer whales was Dr. Michael Bigg. In the 1970s he was employed by the Pacific Biological Station in Nanaimo, British Columbia. One of his first jobs was to figure out how many killer whales were living in local waters. For over a decade, killer whales in British Columbia and neighboring Washington State had been the target of a live-capture fishery. They were herded into harbors or natural coves and inlets where their captors waited with fishing nets to snare them. Already almost 50 whales had been caught and sold to zoos and aquaria. Among the general public, the consensus was that there were loads of whales, maybe even thousands, and that the live-capture industry was not harming the population. Nobody then was too worried about the harm done to the individual animals either during capture or in captivity. Still, there was enough concern at least among some wildlife managers about the status of the population that a census was launched.

Bigg's first attempt at a census involved the distribution of a questionnaire to fishermen, lighthouse keepers, leisure boaters, and any others at or near the sea. On one day, July 26, 1971, and only on this day, they were asked to report if they had seen any killer whales and if so to give an estimate of how many. More than 500 people replied and when their observations were analyzed, the estimate of the killer whale population came out at 200 to 250 individuals: much lower than the thousands that some believed existed. Based on this, Michael Bigg began a more detailed study, where he used photographs of the whales to establish how many were in each pod.

It may seem like an easy task to count the individuals in a pod of killer whales, but it is almost impossible. All the individuals in a pod are never at the surface at the same time, so most counts overestimate how many there are because the person counting inevitably ends up counting the same individuals several times as they surface in different places. Michael Bigg's idea was to use photos to get a more accurate count. He took a lot of pictures of each group and sat down to count the whales. During this work he realized that not only were some of the individuals easily recognizable, maybe because they had a big notch in the top of their dorsal fin or maybe because they had a big scar across the saddle patch (the grey area behind the dorsal fin), but every individual could be recognized from even small scars and nicks if the pictures were good enough.

This method of individual animal identification is called ID photography, and it was developed almost simultaneously for killer whales by Dr. Bigg as it was for humpback whales by a group of scientists lead by Steven Katona on the east coast of the United States (in which case it is the tail fluke of the whale that is photographed when it dives, each one varying in the detailed outlines of the edge or in the white and black pattern on its underside).[1] Today ID photography is used on a multitude of species, from turtles and elephants to swans and lions.

After discovering the usefulness of ID photography, Bigg and his coworkers Graeme Ellis and Ian MacAskie crisscrossed British Columbian waters and photographed as many killer whales as they could find. They gradually built up a catalog of all the killer whales in the area. A few years later they were joined by John Ford, then a young student, and between the four of them a new understanding of the family life of killer whales emerged.

Knowing individual animals is crucial to understanding what kind of bonds exist in a group of animals. Are they related? What are the dynamics between siblings? Between females and males? Between animals that know each other and outsiders? Even such simple questions as how long an animal lives and what role it has in a society are hard to answer if you can't recognize individuals from each other.

After learning how to distinguish individual whales, the researchers understood that a pod or group of killer whales is not just a loose assembly of animals that happen to be swimming together. The pods are incredibly stable, and the individuals in them are always closely related. The families are as tightly knit as a Sicilian mafia family (or at least how they are portrayed in *The Godfather* movies), with the exception of the role of the godfather himself. Instead, killer whales have a matriarch so to speak, because it is the mothers who are the key individuals in the families. The oldest females in a killer whale pod are typically the mothers of two to four other whales, both males and females. The daughters' offspring—the oldest females' granddaughters and grandsons—are also in the pod, and from the day they take their first breaths they will be surrounded by their brothers and sisters, cousins, aunts, uncles, and grandmothers.[2] This was later confirmed with studies done on killer whale genetics performed by Dr. Lance Barrett-Lennard, another excellent scientist that joined the dedicated group of killer whale researchers in Canada.[3]

As in most other animal species, the closest bond between two animals in a group of killer whales is the bond between a mother and her calf. But unlike what is seen in most other species, the bond in the killer whales that Mike Bigg and his colleagues got to know was never broken, neither for the females nor for the males. A killer whale would stay with its mother not only when it was young but for as long as the mother was alive. Even an adult male would stay with the group and continue to live in the closely knit family that it was born into, right next to its mother and its brothers and sisters. New pods were gradually formed when a maternal group grew and smaller family units spent more and more time on their own.

This intimate knowledge of the killer whales' relationships and behavior also led to a completely new understanding of their diet and their role in the ecosystem. It was already known that killer whales would hunt other marine mammals, like seals, porpoises, and even larger whales—Eschricht's famous autopsy had reinforced the impression that they were voracious killers. It was also known that killer whales would eat fish as well; in the Pacific Northwest they had been seen eating salmon.

But studying the whales individually revealed that killer whales who ate other marine mammals were specialists in this diet. They never went for the salmon but focused exclusively on marine mammals. It requires practice to kill a large marine mammal, and to attack large whales also requires well-coordinated cooperation. Those skills take a long time to learn and train. And the killer whales observed by researchers hunting salmon were as exclusive in their dietary preference as the marine mammal eaters were. The researchers called the salmon-eating killer whales "residents" because they appeared to be in the area all the time. The marine mammal eaters were labeled "transients" as they seemed to only be present in the local waters now and then.[4]

Today there is a more nuanced insight into the ranges and behaviors of the different types of killer whales and the picture is not so clear-cut, but the names have stuck, at least in the Pacific Northwest. Among scientists it is now more common to talk about different ecotypes and to differentiate killer whales into fish-eating ecotypes and marine mammal–eating ecotypes. There is also a more nuanced picture of the killer whales' social lives. The marine mammal–eating killer whales live in smaller groups, and the pattern of staying in the natal group is not so fixed; some adult whales will leave the group or establish groups of their own.

It was these North American studies that motivated us students, first on the *Old-Bi* and later at the Whale Center, to start comparable studies in Norway. Inspired by the North American studies and on a spontaneous whim, I had applied for a Fulbright Scholarship to study at the University of California in Santa Cruz. To my utter amazement I got it, so when the second whale-watching season came to an end in Andenes, I packed my bags and left for the United States. Meanwhile, Tiu decided to overwinter in Andenes. In the fall, when students, backpackers, and volunteers started to leave for their "real" jobs or studies, she stayed on at the Whale Center. Jenny Burdon, a biology student from Tasmania, decided to stay as well and the two of them prepared for the winter by furnishing a small makeshift office in the center. As the days got shorter and darker, Tiu hunkered down over data that had been collected during the summer months. There had been enough sightings of killer whales that she had some pictures to sort through and possibly make a start identifying individuals.

One evening Tiu was flicking through the news in a local paper when a small article caught her interest. The police in Tysfjord were warning people not to go out in small boats in the fjord. "It is too dangerous," a police spokesman claimed, "there are so many killer whales that people can walk across the fjord on their backs." The warning was seriously meant as a deterrent but had the exact opposite effect on Tiu. She left immediately with Jenny. Tysfjord was south of Narvik on the Norwegian mainland: some island-hops, a few mountain ranges, a couple of ferry rides, and five to six hours of driving on country roads away. Practically next door, in a country as long as Norway.

It was already dark when Tiu and Jenny arrived at the harbor in Tysfjord. Since they didn't know anyone there, they spent the first night shivering in the barren and unheated waiting room at the ferry terminal. The next day, they scouted for whales and a place to stay, and they found both when they met Rudolf, a friendly fisherman who was using his beautiful wooden fishing boat to transport building materials for a tunnel. He invited them to stay overnight on his boat and from there they saw their first killer whales in the fjord. They soon discovered that the policeman's report had some truth in it, because there were whales everywhere.

Rudolf told them that the whales had been there for a while, feeding on herring that had come in to overwinter in the depths of the fjord. Nobody could tell them for sure when exactly the whales had arrived, but they understood that it was a new thing. They later learned that fisheries scientists believed that the entire stock of Norwegian herring had congregated in the fjord that winter. Not that it was a lot of herring. On the contrary, herring stocks had been reduced to the point where so few were left that it was possible for all the herring to gather in just one fjord. Why they had picked Tysfjord, no one seemed to know.

Rudolf also introduced Tiu and Jenny to other people in the area. One of them was Per Ole Lund, a local freight skipper who took an immediate interest in the whales and wanted to know everything about them. Ignited by the buzz created by journalists, photographers, and wildlife enthusiasts who had heard about the "whale invasion" (and had started their own invasion of the area), Per Ole immediately opened a whale-watching operation in the fjord. This started opportunistically with offering boat trips to see the whales whenever enough interested people arrived at the local hotel (which hitherto had been glaringly empty in the wintertime), but in subsequent years his operation would become better organized and started to attract many more visitors to Tysfjord.

Unfortunately, not everybody welcomed Tiu and Jenny with the same friendliness as Rudolf and Per Ole. Afraid for his reputation, Rudolf sometimes told them to duck so this or that fisherman wouldn't see they were on board his vessel, and in the small fishing harbor where he moored his boat, they found out that it worked best if only Tiu spoke with the fishermen, as she spoke Norwegian. With Jenny speaking English, they were immediately suspected of being associated with Greenpeace, and then the fishermen would turn their backs on them.

It took two years after the discovery of the killer whales in Tysfjord before I was able to make it there myself. In the meantime, I had returned from California and become the mother of a little girl. I had stayed home with her when she was still a baby (her father was, of course, Morten, who had enthralled me by taking me out to hear killer whales that grey and windy day in Henningsvær). I had also returned to my university in Denmark to convince a rather doubtful supervisor that it would be possible to collect enough data to find out if killer whales in Norway had dialects. When my daughter, Camilla, was almost two years old, I brought her along to Tysfjord with her aunt as a babysitter.

In Canada, John Ford had discovered that the stable family groups of killer whales each had their own repertoire of distinct calls.[5] Dialects are common in many species, including our own, and are often created by geographical barriers. We are all familiar with the phenomenon that people from different regions in our country speak a different dialect, and if you are an avid birdwatcher (and listener) you probably know that the same is true for many bird species. But there are other types of dialects: variations in how we speak that are upheld not by geography but by group affinity. Sociodemographic characteristics (like economic status or gender or age or ethnicity) can color your language with different tones. These kinds of dialects are sometimes called cultural dialects because they originate in social entities that are not kept separate by geographical barriers.

The variations that Ford found in the repertoire of calls in different killer whale groups had to be cultural since they were made by animals that were in the same area and often even swam together. Ford reasoned that killer whale dialects were transmitted from generation to generation through social learning and that they functioned to strengthen group identity and affinity. This makes killer whales one of only a few animal species (along with sperm whales, humpback whales, some species of dolphins, and the great apes) that can be said to have a culture.[6] I wanted to find out, was this phenomenon exclusive to the Pacific Northwest killer whales, or was it also true on the other side of the globe?

When I arrived at Tysfjord with Camilla and Morten's sister as babysitter, Per Ole and his girlfriend, Anna Lisa, opened their house to us. Everything about Per Ole was big. Big belly, big arms, big smile, big heart. It was not unusual for him to have three or four people overnighting in his house, and he gladly started his car to drive across the mountain to pick someone up who had flown into the airport in Bodø, 125 miles away. Some of them were tourists that he would later take out on his boat, some were guides he had hired to help with the tourists, and some were local friends without a car themselves.

After we had settled into the house, Per Ole served us a dinner of cod tongues, a delicacy in northern Norway. He told us that they were traditionally sold by children, who made extra pocket money by cutting the tongues from the fish before they were hung on the wooden racks to dry. The tongues were huge, like fish cakes. Per Ole laughed when I commented on their size.

"Remember how big our fish are up here!" he said, dredging the tongues in flour before he fried them in generous amounts of butter. The tongues were succulent and tasty, like cod, but firmer and with a slight rubbery texture that took a bit of getting used to.

Over dinner, Per Ole explained that the arrival of the killer whales to Tysfjord had changed many people's lives, including his own.

"There are twice as many customers in the supermarket, and the fellow with the small cabins to rent is having a great season," he exclaimed. "I may have been the first to take tourists out, but I'm not the only one anymore, there are many doing it."

For Per Ole, taking tourists whale-watching was more than just a business and a way to make money. He had thrown himself wholeheartedly into everything around the whales and was actively engaged in the science and data collection as well. Practically everybody who was involved in research or other activities around the killer whales at that time benefited from his generosity. They borrowed his car, slept in his house, used his tools and workshop, ate dinner with him, or went with him on his boat at no cost. He became passionately involved in conservation, too, not just of whales but of the oceans and everything in them, and he was among the first to raise the issue of plastic in Norwegian waters. Despite not having an academic background, he threw himself into fund-raising and designing studies for research into the effects of microplastics on marine ecology. (This book is dedicated to the memory of Per Ole. He passed away in 2017 from a heart attack in the middle of a campaign against plastic in the ocean.)

After the frustrating years in Henningsvær, working in Tysfjord was bliss, even despite the approach of winter. Most days there were hundreds of killer whales just outside the harbor. The high mountains surrounding the fjord meant that the waters were sheltered, so it was possible to work even when the wind whipped up the sea outside the fjord.

Tiu had succeeded in becoming a doctoral student in Norway, but it proved impossible for her to get Norwegian funding for her fieldwork. Her supervisor kept arguing that she should stop the silly killer whale studies and focus on something sensible—by which he meant minke whales, which were still hunted and therefore commercially interesting. Luckily, a grant from the Finnish government enabled her to carry on. She was helped a good deal, too, by her natural charisma and her uncanny ability to attract people to her studies and devote themselves heart and soul to supporting her effort to understand the killer whales. Some of these were locals like Rudolf and Per Ole, others came from far away and stayed with her and the project for years.

One such devotee was Fernando Ugarte, a Mexican biology student who had traveled Europe with some friends and spent the last of his money getting to the Whale Center. As charming as he was penniless, it was impossible to turn him away. He became Tiu's most trusted lieutenant, driving boats, taking ID pictures, and collecting all kinds of data. Fernando ended up staying in Norway for more than 10 years, eventually completing his own study on killer whale behavior there.

My own supervisor's poorly hidden skepticism when I left to collect data for a study of killer whale dialects was not completely unfounded. It was very difficult, taking us several years of making sound recordings (and a couple of years of listening to them as well). It could never have been done without the help of Tiu and Fernando. Fernando collected most of the sound recordings I used, and his and Tiu's knowledge of which individuals belonged to which pods was essential for my study of the call repertoires. If I didn't know which whales we had recorded, the data was useless.

Between the three of us, we gradually gained insights into the private lives of the Norwegian killer whales, but our studies were encumbered by a problem we had never expected to face: often there were simply too many whales. Getting ID pictures of each individual and sorting and tracking which individuals were swimming together was difficult when there were hundreds of whales, sometimes we guessed up to 500, in the fjord at the same time. And Tiu estimated that the Norwegian population at that time numbered more than 1,500 individuals.

Days with only a few groups of killer whales were easier, both for taking ID pictures and making sound recordings, and slowly Tiu and Fernando began to recognize some of the individuals and could assign them to different pods. The picture that emerged was that Norwegian killer whales, like their American cousins, lived in very tight family groups. They also had group-specific dialects. I did not succeed in getting recordings of all the different groups, but data from about 10 pods clearly showed that the groups in Norway, just like the American killer whales, had distinct repertories of calls.[7]

The inordinate number of killer whales attracted many people, even though it was winter and there were only a few hours of light each day. Whale watchers, photographers, television crews, and scientists mingled in the rented cottages and in Per Ole's house. I was thrilled when both Lars Øivind Knutsen and Göran Ehlmé appeared on the scene as well. A couple of years had passed since we had been together on the *Old-Bi* in Henningsvær, but the experiences there had triggered a desire in them for more, just like they had for Tiu and me. They both came to work on film projects—Lars Øivind mostly above water, Göran underwater—and we spent lots of time together in small boats helping each other out with our different projects.

Working long hours in the field provided interesting observations of the whales' behavior and social organization, opening a window into their intimate lives. One day, I was out on the water with Lars Øivind and Göran. Just as it was getting late and darkness started to creep in, we came across a small group of killer whales. They seemed to be in no hurry, first swimming slowly toward the coast and then changing course and swimming back toward the middle of the fjord again. There were maybe 10 to 12 animals, four of them adult males with almost identical tall triangular fins. (Maybe they were brothers, they looked so alike.) The other whales were a mixture of adult females and younger whales of different ages. It was too dark to continue filming, but we stayed with them for a while. The sea and sky were so beautiful, and there was something magical about slowly following a group of animals that were in no hurry, just allowing the time to pass.

Then, between the adult animals, we noticed a tiny little wrinkled calf struggling to keep up. The white parts of newborn killer whales are often almost orange or yellow, but this one appeared more brownish. And it was so tiny—much smaller than other calves we had seen. Usually, a killer whale calf is almost glued to its mother's side, coming up with her every time she takes a breath, but this calf didn't seem to be with any specific adult. It approached different whales, but it kept falling behind them, and every time they moved on, the calf was left looking alone in the wide expanses of the fjord. The few times we were able to get a closer look at the calf, we saw what looked like scratches or wounds on the side of its body, and even something that may have been tooth rake marks. It was heartbreaking to see the scrawny little calf trying so hard to unite with the other whales, and we could only imagine how lost and hungry it probably felt. Around us it was getting really dark and eventually we had to give up following the group. We left with heavy hearts, thinking that the calf's chance of survival was slim.

The next morning, we went out again more or less to the same place in the fjord where we had left the group the day before. Past midday we found them at the mouth of the fjord. The group was easily recognizable with its four handsome males. We stayed with them for several hours, which wasn't difficult as their behavior was much the same as the day before. We kept looking out for the tiny killer whale calf, but it was nowhere to be seen. Either it was already dead, or it had been left by the rest of the group.

We were never able to confirm who the mother of this little calf was, as it seemed not to have established the close bond that characterizes a healthy relationship. In all mammals, there are cases where the crucial tie between a mother and her offspring doesn't form. Most of these cases lead to the infant's death. Sometimes it is due to the mother's inexperience; sometimes the newborn is sickly or malformed, leading the mother to leave it. Since the calf was so tiny, we suspected that it may have been born prematurely, which may have been the reason for the lack of bonding between it and its family. But the wounds and scratches on the body of the calf were peculiar. Had it been actively chased away by other members of the pod? Did they not recognize it as one of their own because it failed to give the right signals to elicit maternal behavior, and had this led to harassment, even aggressive behavior by some of the older animals? We could only speculate.

The episode with the tiny calf was unusual. It was much more common to see evidence of the strong bonds between mother and calf, sometimes even expressed after the death of an infant. Several times we observed a mother or another family member carrying the limp body of a dead calf around in her mouth, often for a long time.

Astrid van Ginneken, who also worked as a researcher in Tysfjord for a couple of field seasons, documented a particularly gripping episode: "*It was a dark and gray day when we set out. We soon sighted whales and I cheered as I saw a male push a little orange calf to the surface. It took a few minutes for us to realize that the male was pushing a little dead calf around. It was a sad sight.*"[8]

Astrid noted that the male was accompanied by several females and some juveniles. They stayed close together and moved slowly. "*I felt I was witnessing mourning orcas, sharing the grief over the loss of their little family member*," Astrid continued. "*Quietly, we saw the male surface, and at times we saw him push the calf in front of him. The other whales swam at his side. We watched them, hardly moving, to see what happened.*"

After a while the whales started to spy-hop around the boat—vertically poking their heads above the water's surface to get a look around—but apparently not taking any notice of the observers. One of the females spy-hopped and was immediately followed by the male. When he rose up, they saw the little calf slip away from his mouth and slide back into the water. The whales spy-hopped many times and it looked as though some whales were joining the group while others were leaving. Astrid wrote:

> *The remaining whales made several surfacings and turns in unison, seven whales breaking the surface simultaneously. They formed almost a circle with heads turned inwards. Then after one or two dives, they suddenly lined up again and surfaced in one row. Again, the whales came close together and were all directed to something in the center. Suddenly, the little dead baby orca was visible in a flash. Apparently, the whales were around the little calf and taking part in a mutual activity of nudging the baby.*

After several hours of observations, the whales broke up into several smaller groups and left the area. "*We had the feeling we had witnessed something that seemed to be some sort of ritual,*" Astrid concluded.

In 2018, many years after the incident that Astrid observed with the dead killer whale calf in Tysfjord, and far away in the Pacific Northwest, an adult female killer whale was seen carrying her dead calf around on her head or in her mouth for 17 days. It was a disturbing sight and the pictures of the mother pushing the dead calf and swimming more than 1,000 miles carrying it, while struggling to keep up with the rest of her pod, captivated a global audience. Many people followed the drama until the sad end, when she finally gave up and let the body go. Although scientists say that the length of time she carried the dead calf was (probably) unprecedented, it is also generally agreed that this type of behavior is not uncommon. Even if we can never know what an animal is feeling or thinking, this behavior shares many of the same characteristics that we associate with grief in our own species.

These deeply emotional stories resonate with us because we see something in wild animals that we immediately recognize in ourselves, whether it is the grief seen here or other emotions and behaviors like play, curiosity, siblings "babysitting" a calf when the mother is diving, and also frustration and anger. The famous Austrian behavioral biologist (and Nobel laureate) Konrad Lorenz once wrote that we laugh at animals when they do something that reminds us of ourselves.[9] Perhaps, when we witness animals behaving in apparently sentimental ways, we feel a strong connection because we realize that the divide between us and other animals isn't so great after all.

The first killer whales in captivity, which were caught and displayed in the Pacific Northwest in the 1960s and 1970s, are often credited with dispelling local people's fear and hostility toward the animals. Instead of being the bloodthirsty monsters that everyone had imagined, the killer whales proved to be playful and clever. These unfortunate first captive individuals undoubtedly played a big role in changing the perception of killer whales, but the awareness generated by the insights into killer whales' social lives from the long-term studies initiated by Mike Bigg and others was important, too. Similarly, Jane Goodall believes that knowledge about primates' social lives and the realization that they have personalities, minds, and emotions have been critical in creating the profound change of attitude there has been in the way we view our closest kin.

A change in attitudes to killer whales didn't come as easily in Norway as we had hoped. When the local tourist center in Tysfjord started arranging an annual nature festival, they asked us biologists to give talks and photographers and cameramen like Göran and Lars Øivind to show their footage. The international whale watchers participated enthusiastically in these events. A handful of locals came as well, undoubtedly nudged by Per Ole into coming. Money talked, too. Owners of local businesses that were now thriving because of the influx of tourists and nature lovers of all kinds also wanted to hear about the killer whales.

Tiu's growing popularity among the fishermen pulled in a few more supporters. With her quiet manners and genuine interest in the fishermen's experiences and points of view, she had won some over to her cause, though most of them still stayed away. The fishing was not good. There were almost no herring, and the fishing quotas were accordingly small. The fishermen blamed the killer whales and the local association of whalers eagerly stepped up to support the fishermen. "*The time has come to consider an unlimited catch of killer whales*," they argued.[10]

Whalers never really liked killer whales anyway.

CHAPTER 3

Blubber Choppers

I once met a hunter in Denmark who gave me a lecture on hunting ethics. He took pride in doing things properly and in being part of a community where the hunters shared responsibility for preserving what he called the "hunting grounds" (what the rest of us nonhunters would call meadows and woodland). Shockingly, to me, good hunting ethics to him included the indiscriminate shooting of "black birds," that is, birds belonging to the corvid family, such as crows, rooks, magpies, and even the occasional raven. I must have looked appalled because he felt the need to explain himself.

"There are too many crows. They need to be controlled, otherwise they'll kill all the wild birds," was his reasoning (overlooking the fact that the corvids are wild birds, too, and that most of the birds that the hunters shoot—primarily ducks and pheasants reared in captivity and released prior to the hunting season—are not).

His argument, that natural populations of an unwanted species need to be controlled, is regrettably common. It rears its head not only in Denmark in relation to corvids, but everywhere humans see an animal as competition for a resource they want access to. It is the same kind of argument the whalers in northern Norway used when they justified hunting killer whales in the name of protecting the herring stocks.

Until the end of the nineteenth century, most whaling was for slow-swimming species, such as the right whales (which is the name given them by grateful whalers) and sperm whales, which stay almost motionless at the surface when they come up from the depths to breathe. Whaling back then was performed from sluggish sailing vessels that had to lower rowing boats with oarsmen and harpooners to do the actual killing when they encountered whales. It was slow and cumbersome and incredibly dangerous, but still the whalers managed to wipe out almost all the right whales and a good deal of the sperm whales. The northern right whales have never recovered; their numbers are still in the low hundreds. Extinction for them is right around the corner, maybe just a few decades away. The bowhead whale, also a species of right whale, was saved from extinction by virtue of living in the high Arctic, an unforgiving place to work, even for the toughest of whalers. The southern right whales dodged destruction because they live in the southern hemisphere, which was not quite as fully exploited early on. The time of indiscriminate slaughter would come to the southern oceans, too, but by then the target had swung to the rorquals.

Rorquals is the collective name for the fast baleen whales, including the blue whale, fin whale, sei whale, and minke whale. Sleek and rapid like beautifully designed racing cars, they were impossible for sailing vessels to catch up with—and, of course, completely out of reach for the harpooners on rowing boats. Consequently, they were out of reach to whalers for centuries. This changed in the 1860s when the Norwegian whaler Svend Foyn began using a steamship for whaling. When he also invented a harpoon cannon armed with explosive grenades and mounted it in the bow of his ships, the slaughter of large whales increased dramatically. There was no longer any need to put men at risk in the battle with the giants. The whales could be harpooned from the deck of the ship and collected when they had died.

During the first decades after Foyn's inventions, whaling intensified in Norway, and the trend quickly spread to other areas of the North Atlantic and, some decades later, to the North Pacific. By 1910, whaling in the northern hemisphere was already declining. It was getting harder to find the whales, but ships were bigger and more powerful now, so the whalers simply steamed south, to the southern seas and as far as Antarctica, where they systematically wiped out population after population of almost all the big whale species. Most whaling ships hired a majority of their crew in Norway; they were the best gunners and they had almost monopolized the market. Thus, a bountiful and seemingly endless natural resource turned out not to be inexhaustible after all. This grim chapter in the chronicles of man's exploitation of nature is an important part of Norwegian history, both in terms of its economic impact and of the cultural legacy it left. To this day, most communities in northern Norway still have ties to whaling in one way or another.

Andenes is no exception, despite the fact that it was never a real whaling town itself. Just a few hundred meters from the quiet residential street named in honor of Svend Foyn, the Whale Center had become established, with a very different agenda. Right from the start, the center's philosophy had been to build bridges of understanding between different stakeholders, trying to balance the antiwhaling views of most visitors and outsiders with the support for whaling expressed by many local residents. In 1982 the International Whaling Commission had agreed to a moratorium on all commercial whaling, but Norway had filed an objection, which allowed the country to continue hunting the smallest species of baleen whale, the minke whale. This put the Whale Center in a tough spot.

Meanwhile, I had slowly moved up the ranks from galley cook to become one of the marine mammal biologists in residence, more or less by both refusing to leave and working with very little pay. In the beginning of the 1990s, I had finally completed my university studies and was working in Andenes regularly from early spring to late fall, when we would close the Whale Center just in time for me to go to Tysfjord for the winter season.

One sunny summer's day, when I saw a whaling boat coming into the harbor, I decided to pay it a courtesy visit in the spirit of our bridge-building philosophy. Whaling boats in Norway are not like the huge factory ships that went whaling in the Southern Ocean around Antarctica; they look very like ordinary fishing boats of the region with a boom for lifting the nets aboard, and they can very easily switch back to fishing as the season suits. Some have beautiful old wooden hulls like the older fishing boats, though this one was a steel boat, painted blue and white. I knew its business immediately from the conspicuous crow's nest fixed high in the mast. You don't need a lookout to spot fish.

I was especially interested in talking to the whalers about killer whales. It was not a species they hunted (anymore), but I knew that it was a species they had strong opinions about. I grabbed some booklets and brochures from the Whale Center and walked resolutely down to the quayside. I soon picked out the boat with the crow's nest. A harpoon cannon mounted on the front deck confirmed it as the whaler. There was no one to be seen on deck so I climbed on board and guardedly opened the door to the wheelhouse. Inside, screens and instruments were glowing green and orange, and there was a humming noise from some machinery. There was no one around, but I heard muffled voices rising from below deck. I called out and a man's head popped up in a stairway.

"Hi," the man said. "Who are you?"

A little apprehensively, I introduced myself and invited him and his crew to visit the Whale Center.

"I can't come now," the man replied, "we're cooking dinner." He hesitated for a moment, but then invited me below deck instead.

I followed him down the steep companionway, careful not to slip on the steps. Below deck in the spacious saloon, he introduced himself as the skipper of the boat. Another guy sitting at the table with a newspaper was the deckhand. There are three more crew, the skipper explained, but two had gone into town for some shopping and one was sleeping.

The skipper himself was a tall, clean-shaven guy, wearing a black t-shirt and a dish towel around his waist that served as an apron. Perhaps he lifted weights in his free time; his t-shirt was tight around his bulging upper arms. He invited me to sit down and offered me coffee. Then he added with a grin, "You're also welcome to join us for dinner. We're having whale steaks."

I could tell. A huge block of almost black meat was lying on the worktop in the galley. I declined that offer but accepted the coffee. The other guy lowered his newspaper and made room for me at the bench. They were friendly, but skeptical, when I talked about the activities of the Whale Center, and "building bridges of understanding" didn't come easily. We talked about whaling and whales. To them, whales were there to be "harvested"—a word that seemed to imply that they are something you can grow like potatoes in field. As if to underline his opinions, the skipper cut a big slab of whale meat and started frying it.

There was only room for one steak at a time in the pan. When it was done, which didn't take many minutes, he slid it onto a plate and scooped some fried onions from a pot on the back of the stove. Dinner was ready. The steak was bigger than the plate, and there was no room left for potatoes or vegetables. Clearly, no one was missing them. He set the plate in front of the deckhand and then prepared another one for himself.

While they ate their steaks, we talked about the different whale species in the area. The whalers don't care much about sperm whales, the species that we most often see on whale-watching trips, since they are no longer hunted in Norway. Killer whales, on the other hand, they do care about. The Norwegian government's protection of killer whales only a few years earlier was unnecessary, even damaging, the skipper claims.

"They should be removed," the skipper argues, "they prevent the herring stocks from bouncing back." His concern for the recovery of the herring is not his only issue with killer whales. A bigger problem for him and his business is that the mere presence of killer whales frightens other whales away, he claims—whales that he wants to hunt. "Killer whales are like a fox in a chicken coop," he says. "They either create havoc or they eat up everything."

The deckhand offers his opinion too.

"They attack the larger whales," he says, shaking his head in disapproval. "They bite pieces right off their sides."

It is not a coincidence that the name for killer whales in Norwegian is *spekkhogger*, which literally means "blubber chopper." Whalers gave them this name.

I tell them that's not what I've experienced; we sometimes see killer whales and other whale species in the same waters, and it is my impression that most Norwegian killer whales eat mainly herring. I'd brought with me an underwater video of killer whales hunting herring and ask if they want to see it. They agree, and we move so we can see it on the TV they have in the bigger saloon. We watch as the killer whales circle the herring underwater. The fish are packed into a dense shoal, swimming frantically and changing direction as the killer whales, in what appears to be a coordinated attack, swim around them, turning their white undersides toward the herring and emitting air bubbles, which further panic the fish.

The shoal is trapped between the water's surface above and the killer whales encircling underneath. We can hear the melodious underwater vocalizations of the killer whales and the clicking of their echolocation as they move around the fish. As the whales nose into the shoal one at a time, it becomes evident that there is no escape for the herring. The whales slowly orient their bodies into the right position to slap the fish with a powerful stroke of their tails. Underwater, the tail slap sounds like a gunshot. Immediately after the impact, dead or stunned fish pepper the water, and the whales chomp them up one by one.

The men are intrigued by the whales' cunning way of corralling the fish into a tight ball and impressed by how much is going on below the surface, ordinarily out of view to us humans. They comment on the behavior and ask me questions about how we filmed it. Sitting below deck in the small saloon looking at the television, there was at last a brief moment of bridge-building and appreciation for the whales. I can tell why—it's the hunters' admiration for other smart hunters.

There are not many whalers left in Norway now, or anywhere else for that matter, but when whaling was a big industry, killer whales often felt its wrath. It is not difficult to find evidence of this in old logs and diaries from crew members on whaling vessels. One such example was written by the British physician Robert Blackwood Robertson, who joined a British/Norwegian whaling fleet in Antarctica and the sub-Antarctic islands from 1950 to 1951.[1] In his spare time, he scribbled down his experiences and impressions in a book full of detailed observations and colorful tales.

Even as late as the early 1950s, there were still enough whales around Antarctica that the harpooned bodies would be lined up alongside the factory ships, waiting for their turn to be dragged on board, flensed, and boiled. And a dead whale usually sinks. To prevent this, whalers used a hollow spear attached to a hose to pump air into the cadavers to keep them afloat. If they were busy, they might buoy a carcass and leave it floating where it was killed, to be picked up later by the factory ships, sometimes making an easy meal for killer whales. The killer whales began to hear the noises of whaling—the engines, the harpoon guns, the steam winches, and so on—as a signal that dinner would soon be served: an industrial-sized marine dinner bell. Robertson had no appreciation for this demonstration of the killer whales' intelligence and fast learning abilities; rather, his description of the killer whales has the feel of a horror movie: "*Five killer whales, each twenty to thirty feet long, with huge black fins rising rhythmically from the water as they swam toward the ship, and evil black-and-white snouts broken by malignant fang-filled cavities rising occasionally above the water, advanced upon the meal.*"[2]

He compared the killer whales to hyenas and vultures, which he found equally remorseless and loathsome, and continues, "*This pack was about five hundred yards out when I saw them first. They rose to blow, scarcely rippling the water as they did so, puffed each one once, then slid below the surface again, their horrible sickle-shaped fins following them like a humpback's hump.*"

When the killer whales appear closer and closer with each surfacing, he sees their "*cold, black little eyes,*" and then the killer whales are at their prey, "*pulling off a ton or so at a time,*" he asserts.

The whalers' loathing of killer whales may have been prompted by competition for the whale meat but underneath that ran fear and horror. As the killer whales started eating from the whaler's carcasses, Robertson heard the sharp crack of a rifle and seconds later, he saw one of the killer whales sinking in a whirl of spray and blood. For some crew members on the whaling ships, shooting killer whales was more than a leisure sport; it was morally justified in the same way as the Danish hunter's shooting of the corvids. As is often the case in such conflicts, they are more governed by preconceived ideas than facts, and in both cases the competition over the resource may be exaggerated.

The corvids, especially the ravens, crows and magpies, probably scavenge more than they attack and kill live prey (although they also do that every now and then). Interestingly killer whales may also have scavenged more than they killed, at least when whaling was at its height. Two marine mammal biologists, Hal Whitehead and Randall Reeves suggest that scavenging became more important as a food source for killer whales during the whaling era because of the high availability of carcasses.[3] Of course, there will always be carcasses due to natural mortality, but in the days of whaling, natural mortality for the large baleen whales was minimal compared to the mortality caused by whalers.

When whaling came to an end, the free meals became scarce, and the killer whales that had shifted their diet to exploit this food resource were forced to make a new shift. Whitehead and Reeves suggest that after the end of whaling, killer whales may have increasingly turned to smaller prey like seals, sea lions, minke whales, and even sea otters.

Where most whalers only saw conflict—namely that they and the killer whales both took in big whales—a few whalers saw a possibility. From 1908 to 1913, the American explorer and naturalist Roy Chapman Andrews observed whaling operations on board both American and Japanese whaling ships, and he witnessed several times how whalers took advantage of killer whales. Andrews was sent out by the American Museum of Natural History in New York, a museum he later became the director of. He was a dedicated and fearless explorer who went all over the world in search of adventure, dinosaur bones, early human remains, and anything else dead or alive that crossed his path, much in the manner of Indiana Jones (a character who some claim was inspired by or based on his life).

Andrews was a skilled zoologist and a keen observer. In his book *Whale Hunting with Gun and Camera*, he describes how whalers made an effort to get gray whales, called devilfish, to stampede, much like a herd of cattle that get frightened and make a run for it:

> *If three or four ships are near each other when a school of devilfish are found, they draw together, each vessel going at full speed, while the sailors beat tin pans and make as much noise as possible. The whales at once dive, but as soon as they rise to spout the vessels rush at them again. The devilfish go down once more but do not stay under long, ascending at shorter and shorter intervals until finally they are plowing along at the surface.*[4]

However, Andrews adds, "*It is not always possible to stampede a herd, for often the whales will disappear at the first sound and not rise again until a long distance away.*" This is where the killer whales came in as the whalers' helpers. If killers were about, Andrews noticed, it was much easier for the ships to stampede a herd of gray whales.

Andrews also observed that killer whales had a special liking for the tongue of big whales:

> *The killer's habit of forcing open a whale's mouth and eating the tongue from the living animal, is an extraordinary method of attack which has long been recorded by whalemen who hunted the Arctic bowhead. I must confess, however, that I have always been skeptical as to the accuracy of this report until my own experiences with gray whales in Korea, where its truth was clearly demonstrated.*[5]

Sometimes this is all the killer whales want, he says in the book, and they leave the injured whale for the whalers to finish. Other times they snatch the tongue from a whale that has been killed by the whalers but leave the rest of the carcass. Bizarre and cold-blooded as it may sound, the killer whales' preference for the tongues has since been confirmed by other observers.

Andrews also reported that whalers would follow killer whales in the hope that they would lead them to larger whales. The Korean gray whales, he wrote, were so scared of killer whales that they swam into shallow water where they tried to hide among kelp or behind rocks. The whalers knew how to take advantage of this and would follow the killer whales to get an easy catch of gray whales:

> *The orcas are not afraid of the ships and will not leave the whales when the vessels arrive, thus giving much assistance to the human hunters. Captain Johnson, of the Rex Maru, brought to the station at Ulsan a gray whale which had been shot in the breast between the fins. He had first seen killers circling about the whale which was lying at the surface, belly up, with the fins outspread. The animal was absolutely paralyzed by fright. The vessel steamed up at half speed and Johnson shot at once, the iron striking the whale between the flippers.*[6]

If killer whales are able to drive other whales to shore or into the arms of whalers, it seems obvious for whalers to exploit this. So why haven't whalers taken advantage of the killer whales more often? Well, maybe they have. But strangely only in a few places and only a few times in history.

One such place is northern Norway, but the account is old and not very detailed. It was written by Niels Knag, a bailiff to the Danish King Christian V. In 1690 he was sent on a mission to Finnmark, Varanger Fjord, and the Kola Peninsula in the northernmost part of Norway. At that time, it was a cold and forbidding province of the kingdom of Denmark (it is still cold, but no longer part of Danish territory). The area was inhabited by a group of indigenous people that Knag referred to as the Finns. He divided them into two groups: the Highland Finns, who lived as reindeer herders and hunters of wild mammals in the tundra, and the Sea Finns, who were coastal and lived from fishing and hunting marine mammals, especially seals but occasionally also whales. Today these indigenous peoples are called Sami and they still live in northern parts of Norway, Sweden, Finland, and Russia. Niels Knag's report was among the first to describe their way of life.

The king didn't send him to do ethnographic studies, though; his task was to collect taxes in this remote place. It was stipulated that the wealthiest of the Sami people should pay their taxes in the furs of otters, martens, reindeer, and bear, and moreover they should also supply 10 buckets of eider down and feathers and 60 feet of rope made from whale or seal skin.

Knag traveled widely and noted his detailed observations carefully. In a large fjord at the northernmost part of Norway, facing eastward toward Russia, he heard of a strange way of hunting whales: "*In the Varangerfjord there is a reef half a mile from the coast, where the water dogs hunt the whales, mostly in the winter.*"[7] He doesn't describe the water dogs in detail, but they are clearly killer whales, which are still common in the area today. They are the only animals that hunt other whales and describing them as dogs matches the observation that they hunt in packs like dogs or wolves. His next descriptions leave no doubt.

> *When the tide is falling, the hunted whale has to stay inside the reef and get beached on land. The water dogs bite big chunks of the blubber off the whale, and when they get to the meat the whale roars and it can be heard far away. When the Finns hear the whale's roar, whether by day or night, they gather and kill him.*[8]

Knag adds an interesting note that suggests that this way of hunting whales was a traditional and established method, not just a one-off. According to Knag, the Sami in the area each had to pay an old Sami man, named Levi, half a pound of fish a year. Evidently Levi was a shaman, because when he got his fee, it was in his power to make the water dogs herd the large whales into the fjord. Otherwise it didn't happen, Knag asserts. He adds that the Sami in the area ate the whale meat and blubber fresh, though sometimes also salted.

Shamanic powers or not, it's clear that the Sami derived significant benefit from the killer whales chasing their prey into shallower waters. Perhaps it was not always a reliable method of hunting large whales, but it does show that the Sami saw the killer whales as helpers, not as enemies.

Knag's account is old and the veracity of it can be questioned, but there is another more recent account that offers a lot more verifiable detail and is in many ways even more remarkable as an example of collaboration between man and beast. For this, we have to go to the opposite side of the Earth from northern Norway to Twofold Bay in eastern Australia.

CHAPTER 4

The Law of the Tongue

An old man walks slowly down the beach; he limps and leans heavily on two sticks. He looks frail as he walks gingerly between a series of small fires that he has lit on the sand. Every now and then he pauses and slaps the water with the sticks. Sometimes he sings or calls out in a loud voice, "*Ga-ai! ga-ai! ga-ai! Dyundya waggarangga yerrimanran burdyen.*"

Robert Hamilton Mathews, who described the scene in 1904, translated the song to something like, "Hey! Throw that fish up to me on the shore!"[1]

Mathews was a self-taught anthropologist who collected stories about Aboriginal traditions in Australia. The limping, the small fires, and the singing, he explained, were all a trick to attract killer whales and to illicit their compassion. If the old man and his tribe were lucky, the killer whales would drive one of the large baleen whales ashore, making it an easy meal for the community.

Old Aboriginals told similar stories. In a newspaper article from 1914, a man called Charlie Adgery remembered his grandfather's connection to killer whales, or *Beowas* as he called them. The word means "kin" or "brother," and it was the Thawa people's belief that the Beowas were reincarnated tribal warriors from the Dreamtime, born again to the sea in the form of killer whales. When the Beowas came inshore, Adgery said, his grandfather would stroke them on their backs with a long spear. He added that the animals enjoyed the contact in "a cat-like fashion."[2]

The Thawa tribe is part of the Yuin nation. Before the European colonizers arrived, the Thawa land stretched along the southeastern coastline of Australia from Twofold Bay in the south to Wollongong in the north, within present-day New South Wales. Twofold Bay is a large bay, neatly split into two smaller bays, facing the ocean and the big whales' migration routes. Here blue whales, right whales, humpbacks, and fin whales passed from their chilly feeding grounds off Antarctica to their winter breeding grounds in warmer waters. There were killer whales, too, the kind that ate marine mammals. At some point, lost in the mists of time, a special relationship had evolved between the Thawa and the killer whales in the bay. Maybe this started because killer whales in their hunt for marine mammals would occasionally drive a whale closer to the coast or even onto the beach, making it possible for the Thawa people to get an easy catch.

With the European's colonization of Australia, whaling quickly followed. The first whaling station in Australia was established in 1818 in Twofold Bay. The white whalers brought their profession's traditional view of killer whales. To them, killer whales were a competitor stealing their legitimate prey.

In the time before commercial whaling, the seas must have been so well stocked with fish and mammals that their bounty would have seemed inexhaustible. With hindsight, we know of course that nothing in nature is inexhaustible, and definitely not the whale populations, but at that time there were whales everywhere. Whaling wasn't yet a big industry, but it was still good business. Whales were hunted for their blubber and their baleen. Baleen, used for example in corsets and collar stays, combined strength and flexibility in a way that was hard to emulate until the invention of plastics. The blubber was melted to make whale oil, which was easy to then transport and sold well. Before petroleum products were available, whale oil was the best material for lamp oil as well as for many other purposes where today we use synthetic oils. The meat, on the other hand, was often dumped; the European whalers had little use for the massive amounts of it.

The colonial whalers immediately saw the potential of Twofold Bay. The place offered good protection from wind and waves, and the waters were a calving ground for several species of whales. During the following centuries, much of the Thawa people's land would be taken by European settlers and many of the Thawa forcibly removed, but at that time, many Thawa people still lived nearby. From the beginning, the Thawa had to figure out how to coexist with the colonizers as they were hired to crew their boats, and through the Thawa shiphands, some of the colonial whalers gradually developed a different view of killer whales. Unlike the whalers, the Thawa revered the killer whales; they were their allies and helpers. In most places where white whalers established business, they ignored local knowledge, but in Twofold Bay, some of them started listening and learning from the Aboriginal people and a truly remarkable relationship, built on trust and collaboration, between killer whales and man was founded.

One of the first reports we have from Twofold Bay is from Oswald Walters Brierly, an English artist. Brierly left England in 1841 to circumnavigate the globe with his friend Benjamin Boyd, a Scottish entrepreneur who would later become infamous as a slaver and financial fraudster. Boyd decided to interrupt his journey around the globe in Australia, to acquire land and set up business there. In Twofold Bay he founded a settlement to service his various enterprises and bombastically named it Boydtown. In 1843 he set up a whaling station there and hired Brierly as manager. Brierly spent his spare time painting, drawing, and talking with the local inhabitants. He was the first person to describe the unique collaboration between the whalers and the killer whales in the bay.

According to Brierly, the whalemen would approach their quarry in small boats rowed by six to eight men, with a harpooner positioned in the bow, ready to throw a harpoon into a whale when the boat got close enough. It was, of course, dangerous. Once harpooned, the big baleen whales could easily pull a boat for hours, sometimes even days. One throw of their giant flukes could wreck the small open boat and fling the whole crew into the waves. Although whaling in Twofold Bay in this respect was similar to coastal whaling in other areas, it was also very unlike whaling anywhere else, because of the collaboration between the whalers and the killer whales there, something which was already well established when Brierly arrived. He recorded in his diary how a pod of killer whales would follow the boats and how they helped to round up the big whales:

> *They [the killer whales] attack the [humpback] whales in packs and seem to enter keenly into the sport, plunging about the [whaling] boat and generally preventing the whale from escaping by confusing and meeting him at every turn. . . . The whalemen of Twofold Bay are very favourably disposed towards the killers and regard it as a good sign when they see a whale "hove to" by these animals because they regard it as an easy prey when assisted by their allies the killers.*[3]

When the harpooned whale was dead, either because of the killer whales' attacks or because of the whalers' harpoons, the whalers secured the sinking carcass with an anchor and a buoy and left it for the killer whales. The killer whales would eat their favorite parts—usually the tongue and the lips—and leave the rest. After a few days of decomposition, the whale carcass would refloat itself, and the whalers would return to tow the whale to land and start the hard and dirty work of stripping off the blubber.

As the whalers were only interested in the blubber, their only loss by letting the killer whales have their part first was that they had to wait a few days before they could get their own. Locally this "agreement" with the killer whales was called "the law of the tongue." But Brierly noticed that not everybody obeyed the law. Some of the whaling crews in the bay would drive the killer whales away after a catch and not allow them their share. "*The killer whales quickly learned who to co-operate with and who to ignore*," Brierly remarked.

When in 1848 Benjamin Boyd's finances collapsed, Brierly left Twofold Bay and Boydtown turned into a ghost town. Whaling continued, but less intensively, while Eden, a recently established town across the bay, grew in size and importance. In the 1860s, another Scottish immigrant, Alexander Davidson, arrived in Twofold Bay and set himself up as a whaler. He bought up Boyd's abandoned equipment at a good price and built a new whaling station, complete with tryworks, boat shed, and workshops, on the south side of the bay at Kiah Inlet. For the next 75 years, the Davidson family's whaling station would dominate the area and become synonymous with the collaboration with killer whales.

There are still remnants of buildings left at the site of the Davidson whaling station. On a chilly November day in 2003 I visited the place, which today is a historic site just a few kilometers from the Ben Boyd National Park (yes, that Ben Boyd) on the southern side of Twofold Bay. I am staying with friends in Sydney and have rented a car to make the drive. After the first terrifying 30 minutes driving out of Sydney, simultaneously navigating the usual spaghetti of a big city's crisscrossing motorways and learning to drive on the left side of the road, I start to enjoy the drive and the backdrop along the road going south. The road is acclaimed for its scenery and occasionally I do see streaks of blue ocean, but it is still a motorway and the traffic is intense.

When I arrive at the Davidson Whaling Station Historic Site many hours later, my legs are sore from sitting for so long time and I am eager to get out, so I park some distance away and walk to the Kiah Inlet, in the way I picture the Davidsons arrived on the occasions when they didn't come by boat. A rain shower has intensified the smell of the eucalyptus trees that stand tall and erect alongside the track, and a couple of Australian magpies are engaged in a complicated exchange of their melodious calls. At the end of the track it is clear why the Davidsons picked this site for their whaling business. The sandy beach slopes gradually into the sea, making it a perfect place for dragging the carcasses of huge dead whales on land. A few rusty iron containers for boiling the blubber are still there, but the smell of whale oil is long gone.

Just a stone's throw from the beach is the Davidsons' cottage, Loch Garrah, a wooden house with corrugated iron roofs. Like a giant flower pot that somebody forgot to plant, a big iron melting pot sits empty outside the cottage. The Davidsons worked from this place starting in the 1860s. Alexander Davidson's son, John, was part of the whaling crew from the beginning, and in 1878 Alexander's grandson George also joined the family business.

I am not eager to get back into the car so I continue the walk on a track to the tower Boyd built in 1847. It's another pleasant bush walk through lush vegetation, maybe an hour away from the whaling station. Built based on drawings made by Brierly, the tower looks medieval, complete with merlons and crenels for archers at the top. Boyd intended it to function as a lighthouse, but the government never approved of it for this purpose. Instead it turned into a lookout for whales, and for this it must have been perfect. It rests on the top of a promontory and has a perfect view of not only the two bays that constitute Twofold Bay but also of the ocean outside the bay.

I don't see any whales while I'm there. Sadly, there are still few whales in these waters after the intense whaling days around Antarctica. But the wind is strong and the ocean blanketed with white caps everywhere; even with more whales around, today would have been a difficult day to spot them.

The Davidsons certainly used the tower as a lookout point as well. When their sentry would spot a passing whale, they'd hurry on foot or horseback to the whaling station to gather the crew. But often it was the killer whales who would spot the other whales first and swim into the station's waters, announcing their arrival with tail slaps or even breaching in the shallow waters outside the Davidsons' property, throwing their whole bodies out of the water and reentering with a loud splash. Whether it was day or night was all the same to the whales.

Astounding as this sounds, it is well documented and remembered by many. "*It sounded like a rifle shot*," remembered Douglas Ireland, a local resident who witnessed it as a young child and was interviewed many years later, in the 1990s, for an ABC documentary.[4]

Then the whalers knew that they had to rush; a big whale was out there, and the "killers" were on it. The killer whale pod would harass the doomed humpback or right whale and keep it from moving on until the whaleboats could get there. On a nighttime hunt, if a boat lost contact with the killer whales, the men would slap their oars on the surface of the water, whereupon the killer whales would turn back to them to guide them to their prey.

The whalemen observed how the killer whales worked cooperatively to defeat the much larger baleen whales. Some would bite its tail or fins while others would try to jump onto the whale's back to block its blowhole and prevent it from breathing. Others still would attack its flanks or mouth. If the big whale tried to escape, the killer whales would block its way and drive it back toward the whalers. "*He worked them around there like a cattle dog that worked the sheep*" is the way an old man from the town remembered and described the behavior of Old Tom, one of the especially helpful killer whales.[5]

The unusual alliance between whalers and killer whales lasted for at least 70 or 80 years. This means that not only did several generations of whalers benefit from it, several generations of killer whales were involved as well, transmitting the necessary behaviors and skills involved in the collaboration through social learning, just like it happened on the human side.

Sometimes it happened that a killer whale got entangled in the line from the harpoon back to the boat. The whaling crews who disentangled the killer whale reported that the whales would lie still in the water and wait for them to free it. Often, they heard a peculiar noise from the whale during the process, which they described like the sound of a cat purring.[6] It may very well have been the whale's echolocation clicks they heard, but nobody at that time knew that toothed whales echolocated or what that sounded like. In any case, it underscores that the whalers were keen observers.

Sometimes a whaler ended up in the sea. The fluke of a panicked or angry baleen whale is a powerful weapon that can easily turn a wooden rowing boat into a pile of sticks. The whalers had a name for this: they called the whale's tail "the hand of God." To end up in the cold water was incredibly dangerous: if the unlucky oarsman didn't drown or get swiped by the whale's fluke, he risked being taken by sharks attracted to the scene by the blood in the water. Often, however, when the whalers of Twofold Bay fell into the water, one of the huge black-and-white killer whales would swim by them until they were helped back into the boat. The whalers were convinced that the killer whales protected them from sharks in the water and some of them even claimed that the killer whales had lifted them up and carried them to safety.

According to a 1930 newspaper article, "*An old whaler used to swear, on his dying oath, that once when he was overboard and was going down for the third time, Hookie (a killer whale) came along, caught him by the shirt front, and held his mouth out of the water until he got his 'second wind.'*"[7]

The whalers had complete trust in the killer whales. "*They loved them, they were their friends and allies,*" George Davidson's niece Alice Otton recalled in an interview in the 1990s.[8] Many contemporaries found the whalers' trust in the ferocious killer whales astonishing, but it is just as surprising that the killer whales trusted the whalers. They are intelligent animals and they had watched other whales being chased and killed by the men in the boats, and still they were so certain of the pact between themselves and the men that they stayed by the side of the whaleboats, even when the harpoons were flying and everything was chaotic, apparently never fearing that the chase could be directed at them as well.

The explanation for the Twofold Bay killer whales' unusually cooperative behavior may be that with the help of the whalers, the killer whales had a better chance of killing prey that they would normally have to give up on. Fierce as they may seem, killer whales normally do not attack the big baleen whales or sperm whales, not even when they are working together in a pack. They will often aim for the calves and probably only attack adult whales if they sense that they are weakened in some way.

Despite the killer whales being safe with the Davidson family whalers, they were not universally loved or respected. In 1901 a killer whale chasing a minke whale into shallow water accidentally landed on the beach beside his prey. George Davidson, seeing this from out in the bay, immediately directed his oarsmen to row as fast as they could to the shore so they could help the killer whale back into the sea. They knew the animal well: it was a male called Jackson, and he had been their hunting companion for many years. To their alarm, they saw a stranger on the beach walking briskly toward the stranded animal, and instead of coming to its aid he brandished a large knife and purposefully stabbed it to death. On seeing this, the men in the boat became so agitated and enraged that the man on the beach, hearing their shouting and yelling, fled the scene. George Davidson decided to force his crew to return to the whaling station, fearing for the life of the man should the crew get hold of him. This episode had consequences beyond the assassin having to leave town in a hurry. For the Thawa people, it was a grave matter. To them, it wasn't just a whale and a hunting partner that had been killed—it was a family member.

On another occasion, a killer whale was accidentally struck by one of the whalers and died. The incident was referred to in a local newspaper article a couple of years after the stabbing:

> *Many years ago an old whaler named Higginbotham (nicknamed Flukey) in throwing the lance, by accident caused the death of a killer. The same night the natives armed themselves with spears, with the intention of taking his life in revenge for what they considered a great crime, and it was only owing to the intervention of some of the more powerful of the tribe that Flukey was allowed to live.*[9]

For the whalers, the episode with the whale that was slaughtered on the beach also had dire consequences. The other killer whales left the area immediately, and the following year only seven returned, whereas previously there had regularly been between 20 and 30 of them in Twofold Bay. The whalers were convinced that the killing was the reason for their disappearance.

For many decades the whale hunt in Eden was quite regular, and although it was never a huge enterprise, in most years it yielded an average of five to ten whales. When the killer whales had eaten their part, the whale carcasses ended up at the whaling station. There, the blubber was melted into oil in the tryworks, which looks like a brick furnace with large iron pots for boiling the blubber down. "*Oil is everywhere—on the rocks, on the water, the ropes are saturated with it; on the floors which are slippery to the danger point, in the vats, tanks, and tubs—the whole place is covered with the valuable fluid,*" wrote the local newspaper in 1906.[10]

The people in Eden had another, very unusual, use for the carcasses of the big baleen whales they dragged to the whaling station. They believed that the carcasses had medicinal properties and that close contact with them could cure rheumatism and a whole array of other ailments. The treatment was called "the Blubber Cure" and a vivid description of the bizarre procedure is given in an account from 1894, where a local businessman who could only walk with crutches asked George Davidson's permission to try the cure:

> *This was soon afforded him, for shortly after a small right whale was killed, by the Messrs. Davidson, and brought into the trying-out works, and a message was sent over to the afflicted person to come over, and as quick as possible, while the body of the creature was still warm. He did so, and then a large square hole was cut through the blubber into the interior of the whale. Into this the sick man was lowered, after taking off all superfluous clothing. Retaining an upright position, with only his head and hands showing out of the oleaginous cavern, the patient remained in the whale's body for an hour, and was then taken out, on account of the ammoniacal fumes arising from the intestines proving too strong to withstand any longer.*[11]

According to the newspaper article where this story appeared, the man could walk without crutches the next day.

The Europeans undoubtedly learned about this treatment from their Aboriginal partners. The anthropologist Robert Hamilton Mathews who first described the Thawa's connection with killer whales also recounted how Thawa tribespeople with rheumatism would go and sit inside the whale once it was cut open. Despite the promising results, though, the cure eventually was abandoned. Maybe the unpleasantness of entering the stinking and rotten carcass was too high a price to pay after all?

The Davidsons themselves seemed to have preferred a slightly less involved way of taking the cure: instead of descending into the actual carcasses, they crawled into the melting pots. In an old black-and-white picture, George Davidson's son Charlie sits in such a pot full of oil while his brother Jack pours another bucketful over his back. Interestingly, whalers in Bermuda seemed to have used this method as well. There, too, whalers with rheumatism were reported to enjoy sitting in the blood and blubber of a freshly killed whale.[12]

Long before ID photography was possible, the whalers of Twofold Bay used the individual marks and characteristics of killer whales' fins to tell them apart. Brierly noticed this already in 1845, and in 1875 a journalist marveled over the whalers' ability to recognize individual whales: "*These 'killers' are well known to the whalers, and until I saw them close I could not believe it was possible to tell one fish from another, but when I saw their peculiar marks, I could well see that it is as easy to pick out your favorite 'killer' as it is to distinguish a bullock or a cow in a herd.*"[13]

By the end of the nineteenth century, more than 20 whales were known individually and named by the whalers. Often they were named after a Thawa whaleman, in accordance with their belief that deceased members of the tribe were reincarnated as killer whales, but some were also named because of obvious physical characteristics, for instance a bent-back fin. Knowing the whales individually allowed the whalers to notice the whales' different behaviors and personalities and how they apparently had different roles during a whale hunt. "*While 'Tom' and 'Humpy' seem to have been outstanding attackers, 'Charlie Adgery' was also known for the great impetuosity of his movement, assaulting the whale from every angle*," a journalist wrote in a local paper in 1930. The article continues:

> *In one chase, the whale-boat was being towed through the water at great rate. Humpy and Tom were racing along, while about half a dozen others scouted round in a wide circle to prevent the escape of the great whale. Old Humpy seemed to be the boss on this occasion, repeatedly rushing at the whale and taking great bites out of it. As soon as the whale wearied and rose to blow, Tom would fling his body onto the blowhole while Humpy rushed in and snatched at the tongue.*[14]

Tom was simultaneously an adored favorite and an annoying nuisance. He was loyal to the whalers and was probably around for more than 40 years. He frequently visited the Davidsons' station to alert the whalers with tail flops and breaches to the presence of nearby whales, and he was always in the group of killer whales that actively helped in the whale hunt. But Tom was also the whale who would mischievously pull lines from the whaling boat and drag it around at his whim. Or sometimes, he'd purposely throw himself on the rope attached to the harpooned whale, weighing it down and making it harder for the whalers to pull it in. And it wasn't only the whale boats he dragged around. Many fishermen in the area had also been taken for a ride by Tom, who had figured out that he could grab the anchor line and drag boats around, leaving the fishermen equally scared and annoyed. One time, Tom grabbed the harpoon line to a stricken humpback whale and was himself towed at high speed by the big baleen whale until the crew finally caught up with both of them again and secured the humpback (which the crew had feared they were going to lose). Another time, George Davidson was at the stern of the boat explaining the names of the different whales to a visitor when Tom spy-hopped just in front of him and grabbed the line that George was holding on to. The rope coiled around George's finger and crushed it so that his fingertip burst.[15]

The last whale in Twofold Bay was caught in the 1920s and there is no eyewitness alive who can tell of the unusual alliance between the whalers and their helpers. But there are still people who remember George Davidson. One of them is his granddaughter from his youngest daughter, Carrie. Her name is Dorothy Hanscomb, and when I talk to her, she is more than happy to share family tales with me. She is an old woman in her eighties but her memory is vivid and her delight in talking about her famous grandfather is evident.

"My granddad was anything but a rough person, not at all the big fearsome whaler you might expect," she laughs. "He tended a lovely garden and often went to the library to get books." She had a seemingly endless supply of stories about the old times with killer whales. On the table was the family album, replete with pictures of the whaling station and the men working there.

"Every so often, my grandfather would tell us about the whaling times," she reminisces. "The killer whales were like dogs to him and the other whalers. Once he ended up out of the boat and the killers stayed close to him the whole time."

Dorothy also tells me about Jack's drowning in 1926. Jack was George Davidson's oldest son. One day he was returning to the whaling station with his family from a visit to Eden. They rowed across the bay in a small dinghy and, not far from the whaling station, the boat capsized in the turbulence of a riptide and washed everybody overboard. Jack's wife and one of their children survived, but Jack and two other small children drowned. Despite intensive search in the area George Davidson was not able to recover the body of his son.

"For a week, Old Tom continued to swim in a particular area," Dorothy tells me. "And that's where Uncle Jack's body eventually was found." She finished the story by saying that on the day of the funeral for Jack and the children, Old Tom accompanied the boat with the coffins and the mourning family as they crossed the bay to the church in Eden.

Jack had drowned in 1926, long before Dorothy was born, so the memory of Old Tom's behavior may be more family lore than exact records. My point in retelling the story is not to document the astounding behavior of a killer whale—which of course, if the story is true, is truly astonishing—but to underline that the story's endurance in the family recollections reveals how they view killer whales and their strong connection to them.

Most stories of wild animals that have formed long-lasting bonds to humans involve animals that have been "adopted" by humans when they were small, or animals that have been regularly fed. There are only a few examples of spontaneous and genuine partnerships that have developed between wild animals and humans. In one of them, a group of bottlenose dolphins in northern Brazil help fishermen catch fish by driving the fish toward the shore, where the fishermen stand waist-deep in the shallows. The fishermen can't see the fish in the murky water and depend on the dolphins to signal, by a tail slap or a roll of the body, when the fish are close enough for them to throw their nets. Using echolocation to detect the fish, the dolphins are not hindered by the murky water, and just like the killer whales in Twofold Bay they probably get an advantage from the cooperation. It is presumed that by using the men with their nets as a barrier they can drive the fish against, the dolphins get an easier catch—and the fishermen certainly do too.[16] As with the killer whales and the whalers in Twofold Bay, the collaboration between the fishermen and the dolphins has been going on for many decades, through several generations of both fishermen and dolphins.

Interestingly, an old man from Twofold Bay described how he learned to fish as a young boy in a somewhat similar way. His name was Guboo Ted Thomas, and he was a Yuin elder and activist. A few years before he died in 2004, he told the story of how his father and grandfather took him to the beach to fish:

And I remember . . . when they took me down to the beach and I looked very hard and I seen that there was a lot of fish out there, about four hundred to five hundred yards, way out. And he said to me, "There's a lot of fish out there, Sonny." And I looked at him. I daren't say to him "Where's your nets?" He had no nets. I didn't say that. If I'd have said that, he'd have hit me over the head with a bundee (stick), so I had to keep quiet! "Oh yeah," I said.

Thomas watched as his father and grandfather walked up and down the beach, looking out to sea, by turns clapping sticks together, slapping the surface of the water, dancing and singing in their Aboriginal language. They warned the bewildered boy that the fish would soon be coming and gave him a long stick to use when the fish arrived. He couldn't see how the stick would be any use for spearing fish as it wasn't sharpened at all. Then the men stopped:

"Right, wait here now, Sonny," he said, "They'll soon be coming." I couldn't see them. And then I watched the waves coming and I seen one wave coming and another wave and I looked, and the next wave was coming and you could see the fish coming in like that. Coming in. And then I felt them, when a breaker went up, it went right up here to my knee and they'd be hitting me there, and when

it went back it left them high and dry on the beach. And old grandfather said, "Go on, Sonny, use your stick!"

While Thomas hit the fish with the stick his grandfather gave him, he noticed a group of dolphins in the water and came to a sudden realization. Not only were the dolphins the ones driving the fish to the shore, but his father and grandfather had been communicating with the dolphins the whole time.

And grandfather walked out into the ocean and he stood there up to his waist high and then the big dolphin come in. And he put his head on his arm and he held him there, just patting him. And grandfather was talking to him in the language and everything. And the dolphin went, "Chi, chi, chi, chi . . . chi chi chi chi," he was talking to grandfather. And grandfather was talking to him in the language and he just put him down and he just went away. Then a wave come in and I saw the wave and he done a cartwheel over, twice a somersault and he splashed and away he went. That's how we used to sing them in and everything so they'd come.[17]

The way Guboo Ted Thomas tells the story, it is rich with detail and accurate observations of dolphin behavior as we also know it today. The part about the dolphin's sounds is wholly realistic; it was probably echolocation clicks they were hearing.

There are other tales of cooperation between dolphins and fishermen from different parts of the world, some even going back to antiquity. But despite the fact that there must have been many opportunities for whalers to establish the kind of relationship the whalers in Twofold Bay did, there are no other such tales about killer whales.

Perhaps whalers have missed out on the opportunity because their customary view of killer whales prevented them from seeing the possibilities of enlisting their support. On the other side of Australia, south of Perth on the western coast, whalers also encountered killer whales. Whaling commenced here about the same time as on the eastern side of Australia, and as in Twofold Bay, the main targets for the whalers were humpback whales and right whales. At Castle Rock Whaling Station, the site manager mentioned several times in his journal from the mid-1840s that killer whales appeared regularly with the larger migrating baleen whales. But he didn't view them as allies—on the contrary. Several times the whalers in Castle Rock came upon killer whales with a freshly killed whale and managed to take the prey away from the killer whales. Other times, they drove the killer whales away from the whales they had harpooned themselves.[18] The Castle Rock whalers never developed a mutually beneficial relationship with the killer whales, as had happened in Twofold Bay and was surely possible elsewhere. Maybe there was no local tradition of veneration for the killer whales as there was with the Thawa in New South Wales to open the eyes of the whalers to the advantages of joining forces.

In Twofold Bay, there were gradually fewer and fewer whales to be found as the years went by. By the end of the nineteenth century, whaling had decimated the big baleen whales, and either the killer whales went to places where there was more food to be found, or they were decimated, too. The people in Eden heard with concern that further north on the Australian coast, killer whales were being targeted by Norwegian whalers, who viewed them as a pest. And for the first time in history—and the only time so far—whalers argued for the protection of a whale species.[19] Not of the big baleen whales they were hunting, of course, but of their allies, the killer whales.

In 1902, the Davidsons petitioned the Australian government to protect the species they trusted and knew was invaluable to their business. But despite their efforts to lobby high-ranking politicians constantly the next many years, it didn't happen. Not until 1978 would

Australia abandon whaling altogether and declare a "total commitment to protect the whale." But by then whaling had already ceased anyway and the killer whales of Twofold Bay were long gone.

By the early 1920s, there were still a few killer whales to be seen in the bay each year, usually two, sometimes three. But one after one, they disappeared, and eventually only Tom was still there. By this time, he had earned the name Old Tom, and he kept coming back to the bay despite the lack of big whales to hunt. Maybe he enjoyed the company of the people, just as they enjoyed his. He would still follow boats across the bay, and he kept company with George Davidson whenever he was out in a boat. Regardless of Old Tom's antics, there was a special bond between him and Davidson, who was even reputed to hand-feed the whale when he caught fish.

The last baleen whale was caught in Twofold Bay in 1929 and the Davidsons switched to grazing sheep and cattle. And finally, one morning, it was the end for Old Tom as well. On September 17, 1930, his body was found drifting in the bay and was brought ashore at Eden. Unlike most other wild animals whose deaths go unnoticed, Old Tom was mourned by the locals and commemorated in obituaries, not only in Eden but all over Australia.[20]

Initially the whale and its skeleton were offered to the Australian Museum in Sydney, but later the townspeople regretted the offer. A meeting was held, and it was decided that Old Tom should stay in Eden. He was their whale; he belonged there.

He is still there. The day after my visit to the Davidson's whaling station I drove back to Eden where I located the Eden Killer Whale Museum. Inside it I find the remains of Old Tom. When it was decided to keep him in Eden, George Davidson cut up his old friend himself. He saved the skeleton but decided not to boil the blubber as he normally would have done, perhaps out of veneration for his old companion. A year after the whale's death, a small museum was erected, where visitors could see the remains of the famous killer whale from Twofold Bay. The building was later replaced with a new and more modern one, but the story of Old Tom and the other killer whales is still the main attraction in the museum—and of course, Old Tom himself. His skeleton is mounted with the mouth half open and the flippers extended to the sides as if he is on a leisurely swim. But even the liveliest mount can't suppress the deadness of a skeleton.

Old Tom was the last of his tribe—the killer whales never returned to Twofold Bay.

CHAPTER 5

War Zone

Flying into Iceland is like landing on a foreign planet. From the window of an airplane, travelers see a barren landscape of surprising colors. Silvery veins of glacial meltwater transverse a terrain of black, yellow, and bronze. A narrow road meanders across a gigantic coal-grey lava field, forming a curve around a small volcano, passing a place where a handful of houses are strewn across the land before it climbs across another ghostly lava field. It looks peaceful, but below the surface, molten lava is just waiting for a place to erupt. There are no trees and hardly anything green.

As we prepare to land in Keflavik, the sea at least looks familiar: it's a deep blue, with whitecaps making a striated pattern across its expanses. I can't see them from the plane, but I know there are killer whales down there somewhere.

It is 2017, and I have come to Iceland to look into a story I have heard several times but has always been sketchily described. It's a horrific story of how killer whales were systematically persecuted and eradicated in Iceland in the 1950s, and I have never understood what the background for the conflict was or why the situation escalated so badly.

At the airport, I rent one of the cars parked in row after row outside the many rental offices, their brightly colored signage contrasting with the surrounding landscape's earthy greys and browns. With a population of only around 350,000 people, Iceland is one of the smallest countries in the world, but the booming tourism industry means that more than 2 million visitors fly in each year. There are more rental cars for tourists than there are cars belonging to Icelanders.

A handful of areas reported problems with killer whales in the 1950s, and I drive to the one that first attempted to deal with the problem. The place is Akranes, a small fishing village on the tip of the peninsula north of Reykjavik and a few hours' drive away through a weathered landscape. The road hugs the coastline and as I drive north the terrain reveals more detail than I saw from the plane. Not all is lava fields. There are green pastures dotted with sheep, and in many places I also spot the island's iconic shaggy ponies in their paddocks. They are small and strong, built for a harsh climate. According to legend, the Norse god Odin rode an eight-legged Icelandic horse with the name of Sleipner. Indeed, when Icelandic ponies run fast, they shift to a gait that is not found in other horse breeds, called a *tølt*, where the legs move one at a time but incredibly rapidly, giving the impression of a horse with more than the usual number of legs.

Before I arrive in Akranes, I cross the Hvalfjörður, meaning Whale Fjord, but despite the name I only see a little flock of eider ducks on the water. Akranes is colorful to the point where I wonder if the local paint shop has just had a sale. Bright blue, red, yellow, and pink houses dot the landscape, plus houses in electric shades of orange, lime green, and violet. I can sense some of the spirit of the Icelandic people here; to hell with the brown and the black, they seem to say.

Akranes is a traditional fishing village with a large number of fishing boats of different sizes tied up in the harbor. Some are small, like I imagine they all were back in the 1950s. Others are big industrial-looking vessels, and I am guessing that there are more out at sea. On this windless day, the perfect reflections of the boats on the mirrorlike water's surface are only disturbed by two boys in a rowing boat armed with fishing rods.

The harbor is tranquil, quite unlike the description I had read of it in old newsletters from the US airbase at Keflavik.[1] Here in Akranes, on September 21, 1954, 79 heavily armed men stepped on board a waiting convoy of fishing vessels. They were not fishermen but soldiers. They wore uniforms and all of them carried semiautomatic M1 rifles. As the vessels left the harbor, the men lined the decks with grim determination; they had come to help the fishermen get rid of the killer whales, and they were not going to let them down.

It didn't take long before they spotted the first whales. With their tall dorsal fins and black-and-white coloration, they were easy to find and appallingly easy targets. As the soldiers lifted their rifles and started shooting, the sound of automatic gunfire barked across the sea. To kill a whale with bullets is not easy, though, and most were shot many times before they died from their wounds. As the day went on and the sky darkened, the sea slowly turned red with the blood of the killer whales.

Animals that were hit but not killed fled, maybe to die later of their injuries. When the men returned to the harbor late in the evening, they congratulated themselves on a good day's work. They estimated that they had slaughtered more than 100 killer whales. The men were infantry from the US Air Force Base in Keflavik, and their mission marked the friendship and support the US military wanted to demonstrate for their Icelandic hosts. All warlords know how important the demonization of the enemy is, and the Icelandic killer whales were not exempted from this: in the air base's newsletters they were described as "vicious" and "voracious."

Before I arrived in Iceland, I had been in contact with Kristjana Vigdís, an archivist at the National Archive of Iceland. She had looked into their files for me and had found hundreds of newspaper articles from the middle of the 1950s. A search for the word *hahyrningur* (killer whale in Icelandic) before 1950 only gave very few hits, in most years none at all, but between 1950 and 1959, articles including hahyrningur suddenly popped up everywhere. I printed out a huge stack of more than 100 articles but was at a loss how to read them. Eight hundred years ago, all Scandinavians spoke Old Norse, a language that is preserved almost intact in modern Icelandic, but since then the other Scandinavian languages have departed so much from the original that Icelandic is incomprehensible to me.

Not to be deterred, I sought help from Torsteinn Helgasson, a friend of a friend fluent in both the Old Norse of his mother's tongue and my watered-down Danish version of the old language. Between us, we pieced together a chronicle of how, at some point in the early 1950s, some of the area's killer whales had developed a new and very dangerous habit—a habit that was devastating to the fishermen and catastrophic for the killer whales.

The first time a newspaper mentioned problems with the *hahyrningur* was in December 1950. In a couple of paragraphs, the article describes how boats from Akranes had their nets so badly wrecked by killer whales that they were unable to go back fishing the day after. One boat had more than 20 of its nets destroyed, and the skipper was considering giving up fishing altogether. The fishing boats at this time were small, and they all used drift nets to fish for herring. The nets were stringed together at the top, sometimes more than 10 nets were tied on the same line.

A drift net catches fish by their gills as they try to pass through or escape from the mesh. When the nets were pulled on board the fishing boats (in 1950 the nets were hauled by hand), all the fishermen had to do was to shake them vigorously and the fish would fall out. Killer whales had apparently also learned that all they had to do was give the net a good shake, then the herring would fall off and they could eat them up one by one.

Herring are fragile fish, too. Friction with the net or with other fish causes their scales to loosen, and the loss of many scales may lead to a fish dying or becoming disoriented—thus another easy snack for a killer whale. Unfortunately for the fishermen, the killer whales' idea of a good shake was to swim right through the nets. This wasn't difficult or dangerous for the whales because back then nets were made of cotton and easily torn. Some nets could be repaired, but many would be completely ruined.

The fishermen in Akranes and other small fishing villages in the same area had been frustrated and desperate for a while. Hordes of killer whales, sometimes allegedly numbering thousands of animals, were eating their catches and destroying their gear. The fishermen complained to the authorities until the issue eventually rose to a level of national concern. The Icelandic government felt powerless against the killer whales' attacks and by 1954 had turned to the nearby US air base for help.

The Americans were more than happy to offer their assistance. In the middle of the Cold War, the presence of a US air base on Icelandic soil was a political hot potato and there was plenty of local opposition to it. Stepping up and solving the problem of the black-and-white menace at sea was an opportunity to demonstrate not just how effective a military solution could be but also to show the Icelandic people that the US military was not above helping ordinary people.

Despite the brutality of the first round, the extermination of the killer whales had not been completed, and it didn't take long before the fishermen reported that the whales were at it again. Once more, the infantrymen prepared to combat the enemy. This time 60 men went out on the firing squads and continued until the sea was again colored red with the blood of dead and dying whales. Many whales were reported to be wounded and left to bleed to death. The military and the fishermen considered the mission a success, so much so that they sent out a self-congratulatory press release about their history-making whale hunt, which made headlines in both American and European media, including *Time* magazine. The *Time* article mentioned Eschricht's discovery of the killer whale that was found with 14 seals and 13 porpoises in the stomach, perhaps in an effort to emphasize the monstrosity of the whales, and indicated that a newsman on-site during the military mission reported, "*It was all very tough on the whales . . . but very good for American-Icelandic relations.*"[2]

The surviving whales scattered, and for a good while the fishermen were happy with the situation. But a year later, in 1955, when the herring came back to the Icelandic shores, more killer whales followed, creating havoc with the nets and eating the fish yet again. Once more, the Icelandic government asked the US air base for assistance. This time, they decided that an all-out attack would be necessary in order to rid the seas of the whale problem.

This time, two US antisubmarine aircraft were detailed for the mission. Each was loaded with depth charges, air-to-sea rockets, and 800 rounds of ammunition for the aircrafts' machine gunners, who were ready to finish off what the bombs and rockets didn't kill. On Monday, October 24, 1955, the pilots and aircrew boarded the planes wearing their usual combat suits. It looked like war—because it was.

When they found the whales, the two aircrafts split up, allowing the first to swoop low above the surface of the water, dropping its depth charges on a large group of whales directly below the airplane. For a few seconds nothing happened, then the charges exploded, and water shot up like fountains from the site. As the first bodies rose to the surface and blood started to flow from the dying animals, the aircraft crew searched for surviving whales. They were fleeing the scene in panic, leaving their dead and dying family members behind. Resolved to finish the job, the aircraft took up the chase and the gunners started shooting at the surviving animals at the surface, methodically killing whatever fleeing whales were at the surface.[3]

After the first aircraft left, the second aircraft, which had been waiting its turn, dropped its bombs on animals still alive in the carnage below. Then they, too, started shooting with the machine guns from the turret on top of the aircraft. The personnel on the planes later estimated that 40 to 50 whales were killed and reported that the rest of the schools were so scattered that it was impossible to recognize any organized pattern of their escape.

When I charted the accounts of fishermen reporting problems with killer whales, a pattern of expansion emerged. The first reports from 1950 were mainly from the area north of Reykjavik, including Akranes, but in the following years the alarm was raised on the southern coast of Reykjanes, too. In 1952, fishermen from Keflavik on the tip of the Suðurnesja Peninsula as well as from Grindavik on the southern coast also reported that killer whales were wrecking their nets and constantly harassing them. Interestingly, though, the problem with killer whales going through drift nets was only reported from this southwestern corner of the country. There was a lively fishery for herring on the northern and eastern coasts of Iceland as well, but these places did not face the same difficulties.

Charging at nets to feed on fish is a learned behavior, not one killer whales are born with. The way this new and easy way of catching herring spread from the area around Akranes in 1950 to include other regions in the following years is a classic example of a specialized behavior that is transferred from an initial group of whales who figured out how to do it to other whales they associated with. In this regard the mauling of nets, however annoying it was to the fishermen, is another fascinating example of a specialized behavior that is culturally transmitted, just like the feeding specialization killer whales exhibit in other parts of the world.

The US bombing raids continued and included these new areas in southern Iceland in their battlefield. Twice more in the fall of 1955 the combat planes dropped their deadly cargo on the killer whales, but still the whales who were left came back again the year after. And so the raids continued in 1956. As soon as whale sightings were reported, the planes took off, dropping bombs and then shooting at whatever was visible on the surface with machine guns. At the end of the herring fishing and whale killing season, the airmen were all awarded medals for good conduct, and the local whaling company threw a cocktail party to honor their contribution to alleviating the "whale problem."

The written records chronicle only one operation in 1957 and after that no more. Did the previous years' raids wipe out all the whales that had learned to feed from fishing nets or did they "unlearn" that behavior after experiencing the dire consequences? Or perhaps it was the gradual switch from drift netting to purse seining that took place in this period that alleviated the problem. A purse seine is a closed fishing net with the catch inside, which cannot be attacked by predators as easily as a drift net. During the 1950s, the old and easy-to-tear cotton nets were also gradually being exchanged with nylon nets, which were much more difficult to rip apart.[4]

In August 1958, a newspaper reported that there were killer whales in the Akranes area again. But something in the fishing industry mindset had changed radically. Unlike how it had been in earlier years, the fishermen now saw the whales' presence as a good omen. It meant there were fish to be caught. Where there are whales, there are fish. The whale wars in Iceland had ended.

In other parts of the world, the whale wars were just about to start. A conflict with killer whales had been brewing for a while in Norway, and in the late 1960s it came to a boil. As in Iceland, the disagreement was over herring. If you have ever had breakfast at a Norwegian hotel, you will have discovered an item that you don't see on breakfast tables in most other countries. Here, herring takes the place of croissants, and you will find them in a multitude of varieties: pickled herring, red herring, herring in tomato sauce, plain white herring with onions, or smoked herring shimmering in metallic colors. Herring has been an integral part of the Norwegian diet for generations. Today, the fishery still has enormous economic importance.

The Norwegian spring-spawning herring was once assessed as the biggest population of any fish species in the world, but that changed abruptly in the middle of the twentieth century. As fishing gear evolved and became more sophisticated, it also became a lot more effective. A couple of inventions changed fishing radically. The power block, invented by a Croatian-born American fisherman, who was tired of the backbreaking job of hauling heavy fishing nets, is a hydraulically powered pulley that takes over the job of dragging a net full of fish onto the boat. It allowed purse seiners to haul much bigger catches and became standard equipment over a few short years, spreading faster in the fishing fleet than a virus in a wet market. The development of sonar (or echo sounders) also made it easier to locate herring schools in the open sea, and when coupled with the power block, fishing was revolutionized. The effect on fish catches in Norway was immediate. In some years in the mid-1950s, staggering amounts of herring were being landed in Norway, more than 1 million metric tons per year, easily doubling annual catches in the years before the power block and sonar.

In the early 1960s, Norwegian fishermen started feeling the effects of overfishing and got a warning of what was soon to come. It suddenly got much more difficult to find and catch the herring. But after a couple of years, the fishery seemed to recover and the fishermen intensified their efforts. By the middle of the 1960s, the landings of herring were as big as ever. And they included everything: not just the adult herring, which had always been the primary target of the fishery, but also small herring, not yet a year old, and subadult herring, those up to four years old.

Then the fishery came to an abrupt halt.

In 1968, the winter herring season started as usual with the fishing boats gathering along the coast, waiting for the fish to arrive. But nothing happened. The herring didn't turn up.

In Kristiansund, a small coastal fishing town in central Norway, more than 450 fishing boats lingered first for weeks, then months. As the winter continued with no herring in sight, some of the fishermen gave up and started to leave. By February, half the boats were gone. Some had left to try their luck with other species of fish, like capelin. Others gave up the fishing season all together and went home. All along the coastline, in hundreds of small fishing villages, the story was the same. The director of the Norwegian Herring Market, Petter Haraldsvik, called for a press conference and announced the gloomy news. He called it a natural catastrophe.[5]

But there was nothing natural about it. It was simply a matter of overfishing, but at that time nobody thought it could be possible for humans to so completely deplete a resource that must have seemed bottomless. When the year ended, the Norwegian fishermen had managed to catch a mere 20,000 tons, compared to more than 1 million tons annually just a few years earlier. It was shockingly little. In 1969, the catch was even less, and then it was over. In 1970, there were no more herring left. One of the biggest fisheries in the world had come to a full stop.[6]

Killer whales had been a target for whalers before the collapse of the herring fishery. They were among the species of whales that Norwegian whalers hunted locally. Since 1938, a license had been required to hunt any kind of whales, and many boats had such a license. For a long time, killer whales were hunted and taken along with other species, especially minke whales, pilot whales, and bottlenose whales. Only minke whale meat was considered good for human consumption, though. The meat from toothed whales, including killer whales, was used for pet food and given to animals on fur farms, especially in England.

When the United Kingdom banned the import of whale meat in 1973, it could have given the killer whales a respite from the whalers—but the simultaneous collapse of the herring fishery changed that. Looking for someone to blame this disaster on, the fishermen pointed at the killer whales: they were the ones eating all the herring and they were always around when the herring arrived at the coast. The fishermen also offered a solution to the problem: get rid of the whales, they said, and the herring will come back. The culling of killer whales took off.

From 1961 to 1971, more than 1,200 killer whales were culled; almost 500 of them in the first two years following the herring collapse. Many of the whales were killed in the area called Møre, about halfway up Norway's coast. The following five years no killer whales were caught in this area. Possibly so many whales had been taken that there were very few left and those may have become very wary of boats. With the problem solved here, the attention turned toward the Lofoten Islands in northern Norway. From 1978 to 1981, 345 killer whales were taken there.[7]

No one knows how many killer whales were left in Norwegian waters after these massive hunts. And no one will ever know how it affected the remaining whales to have their family groups decimated and fragmented in this way. We don't know the size of the killer whale population in this area before the onslaught began, but killing more than 1,200 of them in a decade must have been a severe depletion—which of course was the whole point of the operation. Unfortunately, and predictably, the culling did not have the desired effect on the recovery of the herring stocks. That took a very long time, and they only started to recover after a temporary complete ban on herring fishing, followed by the maintenance of very restrictive quotas for a time.

The nature of the conflict in Norway was different from the one in Iceland. In Iceland, the clash was over the whales' depredation of the

PAUL NICKLEN

PAUL NICKLEN

Anno 1679 d. 27 December Da kom Denne
Fisk Herudi fioren ved voer Fergested Hvor den
blef Skudt af Velbürdig Christen Leefeul til
Steenalt först med en Müsquette Kuule Giennem
Hovedet oc derefter med 9 Llore rende Kuuter udi
Livet ved nagelen inden Den kunde doe oc Bergis
Dens Lengde Vaar 9½ Sellandts alne
oc dens brede Um Kring
6½ Sellandts alne

Kattegattet, 23 Juli 1861

PAUL NICKLEN

PAUL NICKLEN

PAUL NICKLEN

PAUL NICKLEN

PAUL NICKLEN

PAUL NICKLEN

PAUL NICKLEN

PAUL NICKLEN

fishing gear and the resulting destruction of both gear and catch. Although it is impossible to agree with the horrendous methods the fishermen and their American allies felt forced to deploy, it is possible to sympathize with the fishermen and their losses. The small-scale fishermen affected were poor and their losses were substantial, for many probably devastating. A situation like that called for both knowledge and level-headed action, neither of which was present there and then.

It is unfortunately not uncommon for killer whales to take fish from fishing gear. They are clever and quickly find out if they can make a shortcut to a meal. Depredation of fishing gear of many types has been reported in most places where killer whales are common, including the Gulf of Alaska, Canada, Gibraltar, Greenland, Australia, Kamchatka, and Patagonia. Dealing with the problem is still a big challenge for those involved, which are often angry and frustrated fishermen on one side and authorities and environmental agencies on the other side. In Canada, for example, the official recommendations from government agencies are to stop hauling gear when killer whales are around or move to another area without killer whales to prevent them from associating the vessels with free meals.

In Norway, the clash between fishermen and killer whales was not due to depredation but to the presumed competition over a resource. In other words, the fishermen worried that killer whales would eat the herring they were after themselves. All over the world, this type of conflict is common, and the strategy often employed is to cull the wildlife species involved.

Culling of some species is legal, sometimes even state-sponsored. Wolves and coyotes are among the species most often in the crosshairs. In the United States alone, an estimated half million coyotes are killed annually, mostly due to conflicts with farmers and hunters. Farmers worry about their livestock and hunters argue that the number of these predators must be reduced to protect valuable game species, especially deer. In British Columbia, Canada, wolves are culled to protect mountain caribou, an endangered ecotype of the more common woodland caribou. From 2015 to 2020, between 500 and 1,000 wolves were killed in British Columbia. As killer whales in Iceland were shot from aircraft in the 1950s, so are wolves in BC currently shot from aircraft. Snipers in helicopters are led to the pack by individual wolves that have been equipped with radio collars. Wildlife biologists and scientists argue that there is not enough data to justify these culls and that other threats, such as habitat degradation, including clear-cut logging and easier access for predators on roads and tracks compacted by snowmobiles, are more likely to be factors in the decline of caribou.[8]

The illegal shooting of a wolf in plain daylight in my home country, Denmark, in 2018 didn't end the conflict between conservationists and local farmers. On the contrary, it fired up the antagonism. A study a few years later reported that of 35 individually known wolves in Denmark, three had emigrated and crossed the border to Germany, nine were alive somewhere in Denmark, and nine had died either from natural causes or were hit by traffic. But there were 14 wolves unaccounted for. In a country so densely populated as Denmark, the scientists concluded that these wolves were not just hiding in the bushes. They believed some, if not most, of the 14 missing wolves had been killed illegally.[9]

"Shoot, shovel, and shut-up" has become the mantra of those who want to get rid of wolves and get away with it. Unlike Denmark, both Sweden and Norway allow a limited and much debated culling of 20 to 50 animals each year in both countries, but in both places the illegal hunting nevertheless carries on alongside the legal. In a scientific paper, Swedish social anthropologist Åsa Nilsson Dahlström sums up how the wolf conflict in Sweden has become polarized between wolf-haters and wolf-huggers, urban and rural people, hunters and nonhunters, and the indigenous Sami (who herd reindeer) and Swedes.[10] The parallels to how killer whales have been perceived and persecuted are clear. Apparently, apex predators whether on land or in the water tend to generate this kind of conflict.

Culling of marine mammals may be less noticeable and less debated than culling of mammals on land, but it is very common. Hundreds of thousands of seals and sea lions have been killed in Scandinavia, Scotland, Ireland, South Africa, Namibia, Australia, the United States, and Canada as a "precaution" to protect fish populations. The people involved basically use the same argument as the hunter I once met who shot crows to protect the wild birds that he wanted to hunt.

The culling of killer whales is no longer legal in any country, but the culling of other marine mammals still happens despite little to no evidence that it results in more fish. Most scientists agree that overfishing and other unsustainable practices by the fisheries themselves are a more severe threat to fish populations than any marine mammals are.[11]

When we started studying killer whales in Norway, first in Henningsvær and later in Andenes and Tysfjord, the culling of them had ended just a few years earlier. Many, if not all, of the groups we encountered must have lost family members to the whalers' harpoons. When you know how strong the bonds are between individuals in a killer whale group, it is impossible not to think about what the killing of their family members meant to those that were left alive.

The question of animal emotions, including grief, is explored in Jane Goodall's memoir *Through a Window*. Reflecting on her 30 years of chimpanzee research in Gombe Stream in Tanzania, she describes the strong ties between a mother and its offspring, necessary for the young chimpanzee's survival and encompassing many of the same emotions that we recognize in ourselves. When Flo, an old mother of at least five chimpanzees died, her youngest son, Flint, soon followed her. Flint was not an infant; he was eight years old and therefore at an age where he should have done fine even without a mother. But he didn't. Goodall writes,

> *It seemed that [Flint] had no will to survive without her.... Never shall I forget watching as, three days after Flo's death, Flint climbed slowly into a tall tree near the stream [where she had died]. He walked along one of the branches, then stopped and stood motionless, staring down at an empty nest. After about two minutes he turned away and, with the movements of an old man, climbed down, walked a few steps, then lay, wide eyes staring ahead. The nest was one which he and Flo had shared a short while before Flo died.*[12]

Goodall then describes how over the following weeks Flint gradually became more and more lethargic, until three weeks after his mother's death he died too.

In a killer whale family, with its strong social ties, it is very likely that the effect of culling is as devastating for the surviving whales as it was for Flint to lose his mother. I often thought that it was peculiar that the whales we encountered in those early years were not more wary of us. Did they not recognize the danger in getting close? On the contrary, they very often approached our boats, circling them or swimming under them. Evidently, they could tell the difference between boats that might hunt them (which admittedly were a good deal bigger than our small dinghies) and boats that weren't dangerous. Whalers agreed with this observation: if they had harpooned a killer whale in a group, it was impossible for them to approach the same group of killer whales again.

Canadian salmon fishermen have had issues with killer whales, too. In 1961, the fishermen's complaints persuaded the federal Department of Fisheries to install a machine gun on Quadra Island just off Vancouver Island. The site was picked carefully. Killer whales, or *blackfish* as they were called locally, often traveled through the narrow strait between the islands. Where the machine gun was mounted, the strait was just a couple of hundred meters wide, offering a marksman ample opportunity to aim and hit the whales. This gun was never fired though—at least not at the whales—but that doesn't mean that the killer whales were safe.

In 1964, a young biologist, David Hancock, went with his wife to Triangle Island, a remote and rocky place north of Vancouver Island to spend a month there studying seabirds. David, who spent most of his life observing, researching, and documenting the wildlife of western Canada and the Arctic, is now a charming grey-bearded gentleman in his eighties, but he remembers the trip well. With his wife he boarded the *GB Reed*, a Canadian Fisheries research vessel, in Tofino Harbor on the west coast of Vancouver Island for the 200-mile run to the island.

"*We were barely out of the harbor and into the pounding swells when one of the crew shouted that there was a pod of killer whales portside*," he recalls.[13] The crew member charged passed him and dove down into the hold of the boat to extract a machine gun. "*He quickly fitted this onto an already-installed turret on the bow and proceeded to unload about 40 rounds of huge shells into the pod.*" He adds that the experience was all the more shocking as it was a government research vessel. And that wasn't even the end of it. As they neared their destination, the machine gun was pulled out again to kill Steller's sea lions that had a breeding colony on the rocks next to Triangle Island.

"*Nature was the enemy*," David concludes. "*Back then this was normal.*"

The same year that David and his wife went to Triangle Island, however, marked a turning point in the attitude to killer whales in the Pacific Northwest when a small killer whale was harpooned just north of San Juan Island. Incredibly, the whale, nicknamed Moby Doll, survived both the harpooning and being dragged alongside the boat that harpooned it to the docks in Vancouver. It died after a few months of being held captive in a sea pen, but the whale's friendliness and willingness to interact with people was enough to convince the local aquarium of the potential for profit if it would be possible to catch another whale and keep it alive. The rest of the story is well known. It was possible to catch more killer whales, many more, and to keep most of them alive for long enough to become a cash success for the owners of the aquaria that held them. There is no denying that captive killer whales helped change the public attitude to them. Contrary to the belief that they were merciless killers, they turned out to be trainable, smart and attentive.[14]

But David emphasizes another phenomenon that contributed to the shift in attitude. As a young boy, he had witnessed how the bald eagle was facing extinction and it puzzled him that all the eagle nests along the coast in Washington State were empty. A visit to the Blaine Harbor docks just south of the Canadian border gave him the explanation.

"*Every commercial fish boat tied up at the docks—and there were hundreds of them—had a small white bucket with a sloshing mass of talons and bright yellow legs*," he explained. "*The State of Alaska was paying two dollars per pair of bald eagle legs, and a bucketful of bald eagle talons and legs could fund the purchase of gallons of gasoline to drive a fisherman's boat back to Alaska in the spring.*"

It is believed that over 1.1 million eagles were killed for that bounty, which was sparked by Alaskans' view of bald eagles as "*a beautiful thing, but . . . a destroyer of food [that] should be killed wherever found*," as was remarked in an editorial of a local newspaper in 1920.[15] The bounties were terminated in the mid-1950s, but in the 1960s and 1970s an invisible agent was perhaps even more dangerous and threatening the birds everywhere, not only along the coasts where the fishermen patrolled. Like other apex predators, eagles are especially vulnerable to contaminants because they eat at the top of the food chain where these substances accumulate. Agricultural pesticides like DDT were causing birds of prey to lay eggs with shells so thin they would crack when sat upon, leading to a crash of all raptor populations, not just bald eagles.

David credits the resistance to DDT use in America and its eventual banning in 1972 to the outcry created by Rachel Carson's book *Silent Spring*, published in 1962. Moreover, her book, he points out, was instrumental in awakening a new sense of care for and protection of nature in general. It wasn't just something to admire or a backdrop for activities. The well-being of the natural world and all the organisms in it (including killer whales) is fundamental to our own lives. Carson's book marked the birth of environmentalism.

CHAPTER 6

A Turn for the Better

Once upon a time, fishermen and whales were allies. In the Norwegian classic *The King's Mirror*, there is a description of how whales helped fishermen. And the book lists the many different whale species that lived in the seas around Norway, including one kind known as the "fish driver," "*which is perhaps the most useful of all to men; for it drives the herring and all other kinds of fish towards the land from the ocean outside.*"[1]

It is not possible to say for sure which species of whale the fish drivers were. In *The King's Mirror* they are described as "*thirty ells in length, or forty at the very largest.*"[2] The Viking *ell* was the distance from a man's elbow to the tip of his middle finger, about 18 inches, so that suggests a 50-foot whale: bigger than a killer whale.

Most likely the whales in question were minke whales or humpback whales. The writer added that no one was allowed to hunt them "*since they are of such great and constant service to men.*"[3] In Greenland, the minke whale is called *sildepisker* (herring whipper) to this day. Here, too, locals have apparently noticed the whales' ability to drive fish to the shores or the surface, and they named them accordingly.

The belief that whales help fishermen catch fish has persisted for many centuries. The seasonal arrival of whales along the coast, both baleen whales and killer whales, also heralded the arrival of fish for the fishermen of northern Norway. When whaling intensified in the late nineteenth and early twentieth centuries, this belief led to conflicts between fishermen and whalers. In 1903, an actual riot took place in Mehamn in the northernmost part of Norway. Although Mehamn was a tiny village of only about 120 year-round inhabitants, hundreds of fishing boats crowded the harbor each spring waiting for the arrival of the whales and the fish. During the spring of 1903, more than 2,000 fishermen gathered in Mehamn. The same could not be said of the fish.

For each day the fishermen returned empty-handed and hungry to the harbor, their frustration grew and they became angrier and more desperate. The year before it had been the same. The fishermen had a clear idea why the fish didn't come to the shores as they had always done before, and the answer was right in front of them in the middle of the harbor: the whaling station. There had always been whaling in these waters, but since the industrious Svend Foyn had started using steamships and grenade harpoons, the whalers had swept the Barents Sea clean of whales. The fishermen were certain that the whales drove the fish to the shores and that with most of the whales gone, there was no one to herd the fish into their nets anymore.

The night after Lent in the spring of 1903 a party was held in town, and for some reason, the fishermen were denied entry to the festivities. Blood started to boil, and the fishermen lashed out. During the night, more than 700 men broke into the whaling station and started to wreck it. The next day an equal number joined in and completed the demolition job. With their bare hands they destroyed the whole building, leaving only rubble and wreckage.

The Norwegian government hastily sent the army in to restore law and order and protect the whaling stations, both in Mehamn and in other coastal towns, but a year later they voted in the first legislation in the world that protected whales from whalers. No whaling was permitted for the following 10 years in the northern districts of Norway. It was certainly a landmark decision, but it wasn't done out of concern for the whales themselves; rather it was to protect another, more important industry. The whalers continued their business in other places where either there were no fishermen or the fishermen were not quite as militant.

The scarcity of fish continued to be a problem, however, and the belief that whales were the allies of fishermen eventually disappeared. In many places the conviction was gradually replaced with the exact opposite belief: that whales were eating all the fish and were thus the cause of the failing fisheries. It was precisely this response to the herring fishery that led to the persecution of killer whales in Norway and to the animosity our Whale Center team sometimes experienced toward killer whales in the years we worked in Tysfjord. After the fishery finally collapsed in the late 1960s, herring stocks recovered at a snail's pace—so slowly that for a long time researchers and fisheries experts seriously feared that the herring would disappear altogether. It is estimated that it took until 2010 for the stocks to recover to pre-1960s levels.[4]

After about 10 years of studying killer whales in Tysfjord, the time when they could regularly be found in the fjord every day in the wintertime came to an end. In the late 1990s, fewer and fewer whales arrived in the area and many of them didn't go into the fjord at all. Occasionally some could be found scattered outside Tysfjord, in the large expanse of open sea that is called Vestfjord (although it is not actually a fjord) and all the way down to Henningsvær where we had started our search many years earlier. But just like then, it was challenging to work there because the sea between the Norwegian mainland and the Lofoten Islands is vast and the weather is rough outside the protected waters of Tysfjord. From 2000 to 2010, observations of killer whales were sporadic. Sometimes we saw them from the whale-watching boats going out from Andenes in the summer, and there were also reports and observations from different places along the coast at other times of the year. But mostly they evaded us.

In 2010, everything changed again. A few days before Christmas, Espen Bergensen, a nature photographer from Andenes, went out to take pictures of the full moon hanging low and orange in the wintry landscape. With an eye for the scenic and dramatic, he picked a quiet spot outside town with a good view of the Andfjord to the east and the mountain ranges of Senja across the water. Walking along the shoreline looking for the best place to mount his tripod, he saw killer whales swimming just off the rocky beach and then realized that they were everywhere in the fjord. There were bigger whales, too. With their tall blows, huge broad backs, long flippers, and flukes elegantly thrown into the air before each dive, these whales were easily recognizable as humpback whales. In the following days and weeks, lots of people went out in whatever boats they could get their hands on: fishing boats, fast rigid inflatable boats (RIBs), small cabin cruisers, and kayaks all served to give the curious a close encounter with the giants of the sea.

The whales were gorging on herring. Enormous shoals of fast silvery fish tried to get away from the predators by swimming into the shallows, but the whales pursued them all the way into the shore, also giving people driving on the nearby roads the opportunity to follow the unusual spectacle from land. The excitement lasted almost a month, then the fish and the whales gradually left the area. But the following year they returned, earlier and in even greater numbers. It seemed that after a decade of absent fish and whales, they had finally come back, albeit to a new area 100 miles north of Tysfjord. They have since continued to move further northward. From 2014 to 2016, the main activity was no longer in the Andfjord but along the coasts of the island of Senja and in the fjords around Tromsø. Since 2017, the action has moved even further northeast, to Skjervøy, a small island to the north of Tromsø.

The humpbacks' arrival with the killer whales in the Andfjord in 2010 was a surprise. For one thing, it was surprising that the two species seemed to be coexisting rather peacefully. In other parts of the world, killer whales are known to attack humpback whales; most often the humpback whales, despite being much bigger, will back off and swim away if killer whales are around. But in Norway the two species seemed to tolerate each other. The arrival of the humpbacks was also surprising because they had been nearly absent from Norwegian waters for many decades. They gained full protection from whaling in 1966 but by that time they had been slaughtered systematically for half a century and experts feared the move to protect the species had come too late.

It does happen that a species is able to bounce back even after its population has been reduced to a tiny fraction of its former size. One such example is the northern elephant seal, which was persecuted almost to extinction with less than a hundred individuals left around 1900 but is now numbering at an estimated 150,000 individuals. It has also happened with a number of other species, including the Pacific gray whale, the sea otter, and the bald eagle.

But it did not happen for the Atlantic gray whale or the baiji (Chinese river dolphin), both now extinct. And it probably won't happen for the tiny vaquita—a species of porpoise living in the Sea of Cortez in Mexico with maybe less than 10 individuals left at the time of this writing—or to the northern right whale, with currently fewer than 350 individuals swimming the seas.

Humpback whales have a more hopeful future. In most oceans of the world, their populations are currently growing exponentially. The return of the humpbacks to the Norwegian coast is an example of a successful conservation endeavor. On a day where 20, 30, or even 50 humpback whales feed on herring alongside hundreds of killer whales in a Norwegian fjord, it is just possible to begin to imagine how rich and varied the wildlife in these waters once was—and could become again if we allow it to happen.

There is little mystery to both the disappearance of the Norwegian killer whales in the late 1990s and their return in 2010. The killer whales follow the herring, and the herring move. Like migrating wildebeests crossing the African savanna, herring migrate across the watery plains of the North Atlantic, and like big cats following migrating wildebeest, killer whales follow the herring. The mysterious part is what determines where the herring will migrate to. Historical accounts and modern research in northern Norway agree that most often the herring move in close to shore in the winter. Here they stay, conserving energy and fasting, until the spring comes, when they swim to central Norway to spawn. When the spawning is over, the herring leave the coasts and move offshore to the sea between Norway and Iceland, where they feed on the phytoplankton and zooplankton that bloom there in the short Arctic summer. As the autumn approaches, the cycle repeats itself and the herring embark on their annual trip back to the coasts of Norway. Except of course when they don't and for some reason decide to stay offshore in the winter, as they did in the years between 2000 and 2010. What caused this change is not understood and neither are the northward movements of the herring along the Norwegian coastline every winter since 2010.

In 2016, I led an expedition for the Natural History Museum of Denmark to Senja, the large island next to Andøya, which became a hot spot for whale activity in 2015 and 2016. My years of studying whales and working for the Whale Center in Andenes had ended in the late 1990s when I had moved back to Denmark and started working for the Natural History Museum in Copenhagen. I loved being back in Denmark with my family and friends, and the museum was in many ways an extension of the work at the Whale Center. But I sorely missed the whales. On winter days with crisp air and clear skies, the longing to be out at sea looking for blows and black fins would always hit me with a pang. So with the news of the return of the killer whales, I pounced on the opportunity, and with a little bit of persuasion managed to secure the necessary funding from a private Danish foundation to organize an expedition to Norway.

The expedition, which had dual scientific and educational purposes, had two main themes: the unique co-occurrence of killer whales and humpback whales and the fascinating techniques they each employ to catch the herring. The aim of the scientific part of the expedition was to collect data on the whales' movements and to take biopsy samples in collaboration with researchers from both Denmark and Norway. The objective for the educational part of the expedition was to film the whales and make a documentary for the museum about the remarkable coexistence of killer whales and humpback whales in the same waters.

When we arrived at our field site on Senja late in the afternoon in early January, it was already dark, and we spent the evening unpacking several large crates of gear that we had already sent ahead by truck. We carried the gear from the shed, where it had been stored, into the cozy cabins we had rented, and in a few hours, our setup looked like a cross between a refugee camp and an electronics warehouse.

The filmmakers took over one of the cabins and filled it up with tripods, monitors, batteries, chargers, a couple of big drones, and cables all over the floor, tapping the few electrical outlets in the cabin. I wondered if the electrical system of these holiday cabins would handle the whopping demand of all this gear or if we would blow all the fuses. Soon the film guys settled in with coffee and music, meticulously testing all their equipment. They seemed to be completely oblivious to anything else going on around them.

The next cabin was occupied by the dive team, and they seemed determined to rival the camera team in the amount of gear they could scatter around the place. In the midst of the partially unpacked crates sat my old friends Göran Ehlmé and Lars Øivind Knutsen, who I knew well from the early years in Henningsvaer and Tysfjord. They would be the underwater cameramen on our team. Göran had brought two friends of his to the party and introduced me to Paul Nicklen and Cristina Mittermeier. They were here on their own project, but we would share the cabins and work from the same boats.

I couldn't help being starstruck. Paul is a celebrated photographer from the National Geographic Society and has worked in polar regions both in and out of the water for most of his life. Cristina is a photographer, too, but unlike Paul she also thrives in the steaming equatorial jungles, and her work is as often about the people marginalized in the last square meters of wilderness as it is about the wilderness itself. I soon found out that they were both very easygoing, and any nervousness I had in meeting them disappeared. Having worked together many times before, Göran and Paul were celebrating their reunion with bottles of beer and comparisons of the latest in camera technology. Like debris after a wreckage, the floor around them was littered with dry suits, oxygen tanks, regulators, snorkels, masks, and flippers.

The following morning, we rose early. The sky was still dark and the big mountains looming over the little harbor were only just distinguishable from the sky because they were not sprinkled with stars. But it takes time to get ready for a day at sea, and as we trotted back and forth in the snow loading the two RIBs with gear, the sky gradually lightened to a dark blue. The mountains that had seemed to be all black started to take shape—they were powdered with snow and not black after all.

I am going to be driving one of the RIBs so I dress in a warm flotation suit and my old snowmobile boots, which I found many years ago at a dump in Greenland. They are shabby and battered, but unlike ordinary rubber boots, I know they will keep my feet warm for a whole day in an open boat. In the quiet morning, the snow squeaks as I walk down to the floating dock where the RIBs are tied up. The air is cold, instantly freezing my nostrils. In my experience, that means it is below minus 8°C. When I check a thermometer, it reads minus 14°C.

It is mid-morning when the two RIBs finally leave the little harbor. We stay close together in the narrow strait that leads from the harbor, but when we get further out, we separate, agreeing to call each other on the VHF radio if and when we see something.

Finding the whales turns out to be easy. There are many, perhaps hundreds, in Bergsfjorden, a wide fjord that faces the open sea, but not all in the same spot. One group is very close to the shore in a small bay, another can be seen around some rocky islets in the middle of the fjord, and in the far distance blows from other groups rise above the horizon.

We are looking for signs that the whales are feeding because it is usually easier to observe them when they are occupied. One such sign could be lots of seagulls swarming over the water; another is whales milling around and staying in the same spot for a longer time. From the many years of studying killer whales in Tysfjord, we know that corralling herring is a meticulously choreographed effort, in which the killer whales cooperate and coordinate their movements, circling the shoal of fish, flashing their white undersides toward them, and emitting bubbles underwater to keep the herring together in a tight ball. In the 1990s, Tiu Similä and Fernando Ugarte had discovered that killer whales slap the fast-moving herring with a powerful stroke of their flukes, which either kills the fish or stuns them so they can be picked up one by one and eaten.[5] Once the killer whales manage to control a big school of herring, they can stay with it for hours, taking their time to feed at their leisure. An important part of our project is to film this interesting behavior underwater.

We come upon a group of feeding killer whales around the rocks in the middle of the fjord, but we are about to find out that things are not quite as they used to be in Tysfjord. Initially what we see is familiar. Already a flock of seagulls has arrived, eagerly anticipating the moment when there are dead or stunned fish at the surface and they can join in the frenzy. But just as the killer whales start to whack the herring with their tail flukes, humpbacks suddenly appear not very far away. They move in like a motorcycle gang. Shoulder to shoulder, broad-backed and apparently with total confidence in their superior size, they swim toward the arena at high speed. Their blows stand like columns of mist in the cold air; we count at least five or six humpbacks.

Moments before they arrive at the feeding site, the humpback whales disappear under the surface, and we have just started to wonder what on earth happened to them, when hundreds of herring are suddenly jumping out of the water next to our boat. Then the humpback whales shoot up like underwater rockets fired from below in the midst of all the herring. With their gigantic mouths wide open, they scoop up the herring that the killer whales have so meticulously kept together. Fish are flying everywhere and the killer whales, dwarfed by the humpbacks, reluctantly retreat to the periphery.

Our first experience with the humpbacks turns out to be a common phenomenon. Probably the humpback whales can hear the tail slaps when the killer whales start to feed and rush to the site to join the feast. It makes the work difficult for us, because even though the balance of power between the two species is incredibly interesting, the humpback whales' apparent dominance also shortens the duration of the feeding episodes, and obviously, it is potentially dangerous to be in the midst of a group of animals as huge as humpback whales.

People have been diving and snorkeling with killer whales in Norway for many years now without any incidents or accidents, and Göran and Paul have clocked as many hours in the water filming them as anyone. Still, every time they return from a dive, I am relieved to have them back in the boat. When I ask Göran if he is ever afraid, he tells me no.

"It's always a question of reading the animals and not trying to intimidate them," he says, and I think that intimidating a killer whale would be the last thing I would *ever* try to do if I was in the water with one. "The only animals I'm nervous with are walruses. They can be mean, and they're unpredictable, but killer whales really don't care much about humans," Göran continues.

Paul adds that killer whales, and other animals he has worked with like leopard seals, will sometimes make mock charges, zooming fast toward him, sometimes even showing their teeth.

"It's a bit like playing chicken," he says, baring his own teeth in a smile. "They are testing you, but they're not serious." He adds that a flight response may trigger an attack response, so it is good to hold your ground in a confident way. I decide to stay in the boat.

The humpback whales have certainly added a new and different element to this game. One day, late in the afternoon, we finally have a good situation with killer whales feeding on a large school of herring undisturbed by humpback whales. Both Paul and Göran are down filming and there is a lot of activity, with at least 20 killer whales circling the herring and taking turns diving into the ball. We stay a bit away in the boat to give both the whales and our cameramen space, but we follow the action as well as we can in the fading light.

Paul and Göran have been in the water for more than three quarters of an hour when three humpbacks appear out of nowhere, swimming fast at the surface toward the feeding site. When we spot them, they are still a few hundred meters away but—as we have seen on earlier occasions—they dive and disappear right before they get to the center of activity. From the boat we try to warn Paul and Göran, screaming at the top of our lungs, "Humpbacks, humpbacks! They're under you! Get away! Get away!"

But they hear nothing; they are too far away, and they are below the surface where all above-water sounds are indistinct and muffled. So we watch, biting our nails.

For a few seconds I can see the orange tip of Paul's snorkel, and seconds later I spot Göran, too, but then they both go under again, unaware of the danger. An instant later the humpbacks shoot up through the water and our whole field of view is blocked by their gigantic heads with mouths wide open and throat grooves fully extended. The jaws are so wide that a car could fit inside their gape. For a brief moment everything is a blur of whales and water and fish, then the humpbacks slowly sink back into the water and we start looking for Paul and Göran.

To our immense relief we spot their black heads in the middle of the turbulent water, right where the humpback whales have sunk back into the deep. I quickly put the engine into gear and head over to rescue them. But they're fine, and as they haul themselves back into the RIB, they are laughing and talking nonstop, exhilarated and excited, but also confident that the humpbacks were well aware of what was in the water and that the whales had turned away from them in the last seconds to avoid hurting them.

There are stupendous amounts of herring in the fjord. Filming from a drone, we can see masses of fish moving in the water. Against the white sand of the shallows, the shoals of fish look like a gigantic organism constantly morphing into new shapes. We see the whales, too. With their black bodies and long white flippers, the humpbacks move slowly and elegantly through the shoals, the masses of herring parting on either side of them. Seen from above and compared to the humpbacks, the killer whales look tiny, like children's toys in a bathtub. When any of the whales approach the fish, the herring flee, so a whale moving through a herring shoal looks like it is carrying an invisible paintbrush, coloring an area of white in front of itself as the fish flee and the sandy bottom becomes visible from above.

One day, Lars Øivind mounted a camera on the bottom of the ocean in a small sandy bay and left it recording. He retrieved it after an hour and later that evening we gathered around his monitor to see what he had captured. It was mesmerizing to see the herring moving in waves and undulations, dark against the green color of the water. Sometimes they moved like a single organism, while at other times they broke up and became individuals for a few seconds, then they'd hurry back into line again to rejoin their fellows. A few times a killer whale swam past the camera, a dark silhouette against the surface, and once a humpback filled most of the frame. But most of the time it was just the herring.

It was difficult to determine if the herring are being pressed to the shores by the presence of the whales, or if they have chosen to be in the shallows and the whales have just followed them in. Perhaps when we see the fish in the shallows with the whales, we are seeing the same phenomenon that made the fishermen in Mehamn and all the other north Norwegian small fishing towns love the whales. Whether the whales actively drive the fish to the shores or they are merely following them remains an unanswered question for us, but the fishermen's observations back then were not wrong. They move together.

One afternoon near the end of our time in Senja, I dropped Göran off in the water in a small bay. At the surface we could see both killer whales and humpbacks; there seemed to be a lot of activity. But Göran stayed under for only five minutes.

"It's impossible to film," he complained as he climbed back on board. "I can't see a thing; it's one big moving carpet of herring." He explained that as soon as he got 30 to 40 cm below the surface everything was herring, packed tightly in front of his mask. Every now and then the herring were moving as if there was something behind them, but the density of fish was so high that it was impossible to see what it could be.

We imagined that in a situation like this, humpback whales can more or less scoop up the herring continuously and possibly do not benefit from the killer whales herding them. We don't know what the killer whales do. With that much herring, it is not possible for them to corral them into a dense ball where they can slap them with their tails, but maybe they don't need them denser than this to use that technique.

The comeback of the Atlantic herring is another conservation success akin to the recovery of the humpback whales. Very restrictive fishing quotas, and maybe just a little bit of luck, saved the herring from the same destiny that so many other species have met. With the stocks climbing back to almost historic levels, quotas were again raised and fishing boats started to steam back to the fjords from other parts of Norway.

The herring boats are purse seiners, some of them big, and when they haul the nets, more often than not killer whales are hanging out nearby, like dogs outside a butcher's shop waiting to see what might come their way. And when the last bit of the net is finally lifted on board, their turn comes, as there is always a spillover of fish at this point. Sometimes as many as 50 or even more killer whales will circle a fishing boat and feast on the fish. Humpback whales are not above picking from the leftovers either.

Occasionally we also talk to fishermen in the harbor on Senja, but we don't sense the same level of animosity as we did years earlier. Something has shifted since we worked in Tysfjord. The uneasy relationship between killer whales and fishermen has improved noticeably. In Tromsø we hear that there are examples of fishermen taking whale watchers out to look at the whales and joining the tourists in their joy, taking pictures and posting on social media, even throwing a herring or two to the waiting whales. What has caused this transformation? And will it last? To find out I go to Tromsø, the chief city of northern Norway, to meet a guy who maybe knows.

If Vikings still exist, northern Norway is surely the best place to find them. With a few wooden houses scattered between deep fjords and steep mountains, it would be a perfect backdrop for the *Vikings* television series. Certainly, the man waiting next to a four-wheel drive at the marina could step right onto the film set without much preparation. Tall and sturdy, with a reddish beard, bright blue eyes, and a huge grin he could easily pass as a relative of Ragnar Lothbrok.

In fact, he is Audun Rikardsen, professor of arctic marine ecology at UiT, the Arctic University of Norway. I've contacted him to ask about the shift in attitudes toward whales in Arctic Norway. Why do fishermen post pictures of killer whales on Facebook and Instagram now, when not long ago they argued that killer whales were bad for the fishing? Not to mention handfeeding them herring? Have they started to *like* killer whales?

Audun is a researcher studying both fish and whales and he is a local. He was born and grew up in Steigen Municipality in northern Norway, a small community just south of Tysfjord. If anyone understands the complicated dynamics between fishermen and killer whales, I imagine it must be him.

Audun has generously invited me to join his fieldwork. "We can talk on the way," he had said when we arranged the meeting. I drove from Tromsø to meet him near Skulsfjord, where he lives. A lot of snow had fallen in the previous few days and only one lane had been plowed, leaving an icy surface. I drove slowly in the darkness of early morning. Puffy pillows of whiteness evened out the contours of the landscape, making it look otherworldly in the dim light. Just before I arrive at the marina, a moose and its long-legged calf step out onto the narrow road and pass in front of me. Holding my breath, I watch them turn toward me and promenade alongside the car before they disappear into a thicket of birch trees. Had the car window been open, I could have scratched their flanks.

In the marina, Audun busies himself getting the boat ready. The associations with the Viking era do not extend to the boat. It's a small aluminum boat with a massive 150-horsepower engine and a cockpit that is not much more than a windshield. When Audun is done with the preparations, he introduces me to the other two passengers on today's trip, a master's student and a colleague from another department at the university. Along with the introductions, we are each assigned different tasks. I am asked to take ID photographs of the whales we encounter.

As we head out at full speed, I stay outside on deck because I want to take in the incredible scenery—but as the boat slams brutally into the waves, water cascading across the deck, I surrender almost immediately and seek shelter in the crowded cockpit. We are not on a pleasant sightseeing cruise. Audun chats and laughs, but it is impossible to hear anything over the roar of the engine, so I smile and nod, trying to dampen the impact of the rough ride by flexing my knees and clinging to a metal bar in the roof of the cockpit.

We search the bays and fjords we pass on the way, but they are empty of whales. Audun seems to know where they are hiding; he was also out doing fieldwork the day before and has set his course for the same area. It's Eidsfjord, almost an hour's drive on the water from Skulsfjord. When we get there and he finally slows down, it's a relief. I assess the position of my internal organs and loosen my grip on the metal rail. The fjord is as flat as a lake, and we spot a group of killer whales almost immediately.

I don't get a chance to chat to Audun about whales and fish straightaway because we are immediately busy. Audun is armed with a big air gun and a heap of determination, and we spend the whole day taking skin biopsies of killer whales. Even the tiniest skin sample contains DNA that can reveal the genetic affinity of the whale it comes from. Is it related to other whales in the same group? Is it related to other killer whales in northern Norway, or to Icelandic whales? Or to Greenlandic whales, even? How long have these populations been separate from each other and from populations of killer whales elsewhere in the Atlantic and even in other oceans? The DNA in skin samples may shed light on this, and analysis of stable isotopes in blubber biopsies also gives information on levels of contaminants like heavy metals and PCBs as well as confirming their main diet over the last months.

To get the samples, Audun climbs onto the bow where a railing has been fitted, on which he can lean so he can use both arms without falling into the water. We approach the killer whales carefully; it is necessary to get quite close in order to shoot a biopsy dart. I stand behind Audun, ready to take an ID picture of the darted animal. Audun is not only a professor of Arctic biology but also a world-renowned nature photographer, so it is an intimidating task.

The group of killer whales we are following consists of two large males and three to four adult females, two of them with calves of different sizes. Often the males would come up between us and the rest of the group, so they are the first ones targeted. When they surface near the boat, I can tell that Audun is ready to fire. With a loud "POP!" the air gun goes off, and in my viewfinder I see the dart flying over the back of the nearest killer whale and into the water. A miss. I get a picture, but I already have a few of this particular male. We slow the boat down and go back to search for the dart. When we find it floating in the water, the whales are already far away. We start a new approach, and once again Audun fires. This time I see the dart hitting the whale before it breaks off and falls into the water.

"Did you get the ID?" Audun shouts from the bow.

"I think so!" I reply, sincerely hoping that I did. And that the image is in focus.

While we go back to search for the dart, I discreetly take a quick glance at the picture. It looks okay. We get a biopsy from the other male easily enough, but the females take longer. Once a dart is retrieved, Audun's student carefully picks out the small piece of skin with a pair of tweezers and puts it into a vial. Then the sample is registered in the notes along with my picture numbers while Audun prepares a new dart for the next biopsy. In the stillness, when the motor is idle, we hear the sound of the blows from the killer whales across the water. Two ravens croak as they pass overhead. They seem to enjoy the scenery, too, alternating between flying almost in synchrony with their wing tips touching and doing silly aerial acrobatics like flying upside-down or doing daunting barrel rolls.

We continue working as long as there is light. One by one we get samples and ID pictures of all the adult animals in the group. Audun tells me he never targets the calves. The calves are harder to get ID photographs of, and he doesn't want to stress them. Later we find another group travelling along the shoreline, and when we are done with the first animals, we continue with those. Apparently, there are not so many herring in the fjord, based on the observation that none of the groups are feeding, and two humpbacks that we come across at the entrance to the fjord are not approaching the killer whales. Perhaps they know they are not feeding and therefore are not interesting to them.

The light is almost gone when we wrap up the day's work and pack down the cameras and the dart gun. The pink and yellow hues in the sky paint the snow on the mountain in beautiful pastels. There is no vegetation on the mountainside, and its slopes run unhindered into the sea. Where they are too steep for the snow to settle, the grey and black of the rocks contrast with the whites of the surrounding areas. The mountains look exactly what they are. Rock solid. We are not in a hurry to get back. I stay outside on the deck and in the late afternoon there is finally time for some of my questions.

"Yes," Audun agrees, when I ask about the more positive attitudes that we sensed in Senja. "They have changed. It has had an enormous impact that so many people travel up here and that the visitors are so much in awe of what they see. In a way, you can say that it has opened the eyes of local people to something they have not really been aware of before, or at least not as a phenomenon that was extraordinary."

The influx of whale watchers, photographers, and nature lovers of all kinds generates a lot of income in the area and that is obviously important, but it goes beyond that. Audun explains, "People are proud of the whales now. They have realized that what they have here in their own backyard is exceptional. The whales have become icons of this place, just like the midnight sun and the northern lights."

The tensions with fishermen have largely dissolved because there is so much herring, Audun tells me, adding that some fishermen actively follow the whales as they drive the fish to the shore. Near his home in Skulsfjord, where I met him that morning, it is not unusual to see small boats waiting on the water for whales to drive the fish into the shore and then setting their nets around the shoal. Those fishermen obviously do not see whales as their enemy, and neither do the fishermen on the bigger fishing boats. But Audun also stresses that having many fishing boats and whales in the same area still presents a potential risk to both whales and fishermen.

"The killer whales here have quickly learned that a fishing boat means food," he says. "Humpbacks have learned this too." He then describes the same phenomenon that we have also observed around Senja. When the whales hear the winches on board a boat going into action, it signals that a net is about to be hauled.

"I have sometimes seen killer whales abruptly stopping what they are doing to set course for a fishing boat which is several miles away," he says. He explains that there will always be fish that are forced through the mesh or leak out of the net during the hauling process, and the killer whales, especially the adult males, swim incredibly close to the net and the boat to get those fish. After the bulk of the catch has been pumped out of the net into the belly of the boat, the fishermen wash the remaining herring out of the net, kicking off a feeding frenzy where all the killer whales and usually some humpbacks go berserk.

It has happened that killer whales get inside the net, too.

"Once," Audun says, "there was a situation with four killer whales caught in a purse seine net and the skipper called me and asked if I could help." He explains that in a case like that the fishermen are not allowed to just open the net because then all the fish would be released too, and that's illegal. Herring are very fragile, and most of the herring would die if they were dumped. The rule is mainly there to stop fishermen discarding a catch because they think that the fish are too small. So, if a whale is caught, the skipper has to call the fisheries authorities to ask permission to release the whales and the fish.

"When I got to the site where the whales were caught, it was clear that they were stressed," says Audun. Making big circles with his hands, Audun describes how the whales were moving quickly, pacing the circumference of the net, and breathing fast. The rest of the pod was waiting just outside the net, clearly aware of the predicament of their family members trapped inside it. The fishermen were equally stressed about the situation and eager to save the whales, as well as any part of their catch that was still possible to recover. In this case they managed to get all four whales out, but a situation like that doesn't always end happily. A few killer whales have been caught and drowned in other incidents, but Audun is more worried about the humpback whales because they are so big and can more easily get entangled in the nets.

As we approach the little harbor we started from earlier in the day, the sky is a deep blue against the white mountains, and we pass a few other boats. Most of them are different kinds of whale-watching boats going all the way back to Tromsø, and a few are fishing boats that will probably continue working for many hours into the night. As we pass Skulsfjord, he points to a small bay and says, "I was once out taking pictures here when I was contacted by some fishermen who had a problem with killer whales." It turned out that it was a very small fishing boat, too small to haul their catch into the boat, so instead the fishermen gently dragged their purse seine net full of herring to anchor it in the bay and wait for a bigger boat to pump the net. But a group of killer whales was following them, like a pack of wolves stalking a herd of sheep. There were at least 30 to 40. When Audun arrived, the killer whales had started to circle the net, making mock charges and panicking the fish, which were swimming faster and faster in tight formation. All of a sudden, the fish performed a desperate escape maneuver and swam downward, dragging the net with them, along with the floats at the surface that were supposed to keep the net from sinking. It was a free-for-all then. The killer whales swam into the net and gorged on the fish.

The fishermen and Audun tried to scare the killer whales out from small motorboats, but they were powerless against so many. When the frenzy was finally over and the whales began to leave, the fishermen gained control of the net again. They estimated that there were only 20 tons left out of a 250-ton catch. What the killer whales hadn't eaten spilled out of the net and ended up on the bottom of the ocean.

Audun shakes his head. "The fishermen were not happy," he says, "but with so many whales and so much fish, it is inevitable that there will be more incidents like that."

So far, the newfound peace agreement between the killer whales and the fishermen has endured despite the increase in interactions. With proper management of the herring fishery, there is hope that it may endure.

CHAPTER 7

The Whales in the Potato Field

Until late afternoon, June 6, 1981, had been an ordinary day by the shores of Austnesfjorden in the Lofoten Islands. It was cold, maybe just 10°C, but that is not unusual for the beginning of June in this part of Norway. There were clouds around, but the sun was still high in the sky because it was already that time of year when it never sets at these latitudes. It was a calm day, with little wind, and the reflections of the high mountains around the narrow fjord were undulating slowly in the water. A group of maybe 20 to 25 killer whales had been swimming in the middle of the fjord for some hours. They kept going back and forth: perhaps they were hunting for herring, perhaps they were just resting and hanging out. Occasionally their blows could be heard across the water. It was quiet except for the seagulls flying over the water. Their shrieks mixed with the melancholic calls of the curlews guarding their territories on land.

Seeing killer whales here wasn't unusual. The fjord was a good place for herring, and various species of whales would occasionally venture into the bottom of the fjord to hunt them. But late in the afternoon, Kjartan Krane sighted something odd. He had been out the whole day planting potatoes in his field, which ran all the way down to the water's edge. From time to time he had stretched his back, checked on the weather, and glanced at the whales out in the fjord. But now when he looked up, he noticed that some of the killer whales had split from the larger group and were swimming at high speed toward the shore right at the edge of his potato field. To his astonishment, a big male at the front of the group continued into the rocky shallows among the kelp, then slowed down a little bit before launching itself with a sudden jerk right onto the beach.

Kjartan watched dumbfounded as the other whales, one by one, almost as if they were patiently awaiting their turn, followed the male and beached themselves next to each other on the shore. Before he could think of anything to do, 14 huge black-and-white animals were lying on the beach. After a little while, the big male and four of the smaller ones somehow managed to wriggle backward and reenter the water, but the remaining nine animals seemed to be stuck.

Kjartan rushed down to the shore, but he was unsure what to do; they were lying well out of the water, much too far up the beach to get themselves back in. And Kjartan knew the tide was going down. He turned around and ran the other way to look for help.

I learned these details not from Kjartan Krane himself but from his daughter, Sigrund Krane. During the first year I served as cook on the *Old-Bi*, the incident was only a few years old, and still fresh in the mind of many. Since then, Kjartan had passed away, but Sigrund still lived in Austnesfjorden, so naturally, I went there to hear the story from her.

On a fine day with little wind and the sun shining through a thin veil of clouds, I drive to Austnesfjorden to meet up with Sigrund. She lives just a few kilometers from her childhood home and the place where the whales beached themselves. I enjoy the drive. The Lofoten are islands of steep mountains, blue fjords, and an astonishing number of tunnels and bridges connecting the people who live there to the rest of the world. The mountains rise almost vertically from the sea, mostly barren of trees due to the high latitude. They are huge and grey, bordered only at the bottom with a narrow rim of green. The road winds its way along the coast, every now and then rising to climb a rocky outcrop or dipping into a tunnel, and I struggle to keep my eyes on the road and not be too distracted by the magnificent scenery. Over the fjord soars a white-tailed eagle, with wings so broad that local people call them flying doors.

When I had called Sigrund for directions, she'd told me to look for a house in front of a greenhouse on a narrow dirt road shortly after turning off the main road. This is exactly where I find her. Sigrund is a middle-aged woman with strawberry-blond hair. She is wearing a fleece jacket and is accompanied by a very friendly golden retriever. She suggests that we go straight to the place where the stranding happened, telling the dog to stay at the house as she walks to the other side of my car and jumps in. We continue on the narrow road past her house. Small farms and houses are scattered at the foot of the mountains. Both cattle and sheep graze the modest fields that extend almost all the way down to the shoreline. Judging by the sparse grass cover in the hayfield, it is clearly not agriculture at its most efficient, but nevertheless it is astonishing that it is possible to make a living from the land at all, so far north of the Arctic Circle. Sigrund tells me that people who live here combine farming with fishing or work in the bigger towns a bit further away.

When the stranding took place in 1981, it was maybe the first indication of a change in the attitude to killer whales in Norway.[1] The whales didn't just land in a potato field, they landed in the heart of the Lofoten Islands, where the hunt for these whales had been intense in the years just prior to the stranding. For the local people, the easiest thing would have been to kill the whales, but most unusually for a small fishing community in northern Norway at the time, they decided to try and save them instead. It is this remarkable turn of events that I was interested in. What caused these small farmers and fishermen to perform such an extraordinary act?

After a few kilometers' drive Sigrund tells me to pull over, and we get out of the car.

"This is where it happened," she says, as we walk to a small shallow bay. Driftwood, a few remains of fishing gear, an orange buoy, and a polystyrene marker for a fishing net lie scattered on the beach. "The whales came up here, next to these big rocks and they were spread out all along the beach from here to the little shed down there." She points at a small red cabin at the other end of the beach, maybe 100 yards away. She also points out the location of her dad's potato field and her childhood home. It is a small red house on a hill at the end of the field.

"My father came running home, completely out of breath. He yelled at me to hurry up and to bring my new camera." Sigrund had just graduated from college and treated herself to a brand new SLR camera, which she brought back home for the holidays with the family. She hurried to grab the camera and then followed her dad back to the beach with her little brother. On their way they met others already heading the same direction. Some were running, some were on bicycles, and others were coming by car or on a tractor.

Even before a time when everyone had a cell phone in their pocket, most had already heard the gossip and it didn't take long before everyone from the small community was gathered around the animals at the beach. Somebody had called the police and when they arrived, the sight of the cop cruiser yanked the last few people out of their houses. A police car was a rare sight in this place.

On the beach, standing next to the helpless animals, the villagers discussed what to do. It was clear that the whales were unable to get back to the water by themselves. It was equally clear that they were distressed. From time to time the onlookers heard them calling with strange high-pitched sounds. And out in the fjord the five whales that had escaped from the beach were patrolling the shore, swimming slowly, waiting patiently, but unable to do anything.

"There was a lot of discussion going on," Sigrund tells me. Some people considered the possibility of making money off the whales. Most of the men, both the old ones and the young ones, had worked on whaling boats, and this many whales at one time was almost too good to be true. The men, however, soon realized that it was the wrong species. The one the whalers were after was the little minke whale, the smallest of the baleen whales. That was hunted for its meat, but killer whales were never eaten in Norway. At most, the meat could be sold for animal fodder, so they quickly rejected that idea. The policemen had brought rifles and were prepared to kill the whales. But some people argued that it would be a terrible smell with so many animals rotting on the beach, and others felt that it just wasn't the right thing to do.

The whales seemed unharmed. Some of them had scratches and were bleeding from the wounds, but they appeared not to have injured themselves badly in the beaching event. But with the tide still receding, it was clear the animals would die if something was not done quickly. Somehow, during the discussion of what to do, compassion won the day.

"We could hear that the whales that had rolled off the beach by themselves were calling out for them from the water," Sigrund says. "We could see big whales lying next to small calves on the beach." She believed that was deciding factor. When people realized that they were mothers lying there with their young ones, it activated a kindness, an urge to help.

While Sigrund and I have been talking, the wind has picked up, so we decide to go back to her house to continue our discussion. Her dog greets us at the door, but before we go inside, Sigrid shows me her greenhouse. As if to defy the harsh climate outside, it is a surprising little miracle of heat and lush foliage where she nurtures hundreds of colorful flowers. Then we go into her house where she makes coffee and brings a big brown envelope full of newspaper clippings to the table. They are old articles telling the story of the rescue mission, some of them illustrated with the pictures Sigrund took with her new camera.

Sigrund leafs through the photographs, commenting on people she knows in them. The pictures of the whales show the animals scattered along the beach. Some of them were far up on the beach; others were partly in the water, but still too far up to move back by themselves. Many were lying flat on their bellies, others had keeled over and were lying on their sides. Somehow that made them look even more stranded and helpless.

"When we decided to rescue them, that was all that mattered," Sigrund tells me. It became the children's job to keep the whales cool and wet with seawater. They ran home to get buckets, and when they returned to the beach they started making rounds to the water's edge to fill them up and carry them back to the whales. She remembers that the whales sighed audibly when they splashed the water on them.

Pausing from reading the old newspaper clipping, she looks up. "There was some sort of contact, both because they reacted so strongly to the cool water and because we could see their eyes," she says. "I don't know if they sensed that we were there to help. But I think so." It never occurred to them that the killer whales could be dangerous, Sigrund says when I ask if they weren't scared of the huge animals armed with the most impressive set of teeth in the animal kingdom. "We just felt so sorry for them," she says.

While the children were busy with the buckets, some of the men tried to lift the whales off the ground with large poles and roll them back into the water. They started with the smallest whales, but it was incredibly hard work. It was difficult to get a footing on the beach's round boulders covered in slippery kelp. And even the smallest of the whales probably weighed more than half a ton. The men nevertheless succeeded in lifting and rolling the three smallest whales back into the water and far enough out that they could swim. A sense of relief spread among the village people as they looked at them in the water. But it was too early for congratulations. The whales were only in the water for a very short while, then they swam right back to the shore where they beached themselves again.

It was frustrating to see the hard work reversed by the whales swimming back onto the beach, and it led to more discussion among the men involved in the rescue operation. It occurred to some of them that it might have been a mistake to start with the smallest ones, as they were probably calves desperate to be reunited with their mothers. The men, therefore, decided to do the opposite: start with the big ones and hope that they would stay in the water.

After the hard work with the calves, it was obvious that it would be impossible to lift the big whales back to the sea with only the help of the poles; for this, they needed something more powerful. Sigrund's uncle Halvdan had a 50-foot fishing boat, which he now went to the harbor to fetch. The engine of the boat should be strong enough to pull the whales off the beach, they figured. Luckily the place where the whales were beached had deep water coming relatively close to the shore, so when he came back with the boat, he could get close without running aground.

From the boat, Halvdan threw a rope toward land and one of the young men on the shore waded into the cold water to retrieve it. On land, the men helped each other fasten the rope around one of the bigger whales. When they were satisfied that it was fitted well around the whale, they signaled to Halvdan to start pulling. He reversed the boat very slowly while the men on land were prepared to shout if something went wrong and he needed to stop. The rope started to uncoil and the engine noise from the boat was the only sound heard as everyone fell quiet watching the rope gradually grow taut. The whale was motionless, it was impossible to tell if it understood what was going on. Slowly, the rope tightened until it was as straight as a steel wire, and the onlookers, holding their breaths, could see how the boat started to pull on the whale, their eyes going nervously back and forth between the boat and the whale.

At first, the rope just seemed to dig into its tail, but then, its entire body heaving with a massive shudder, the whole whale started ever so slowly to slide over the rocks toward the water. The slippery kelp on the rocks that had made the work so difficult for the men with the poles was now helping make the pulling easier and cushioning the whale from the sharp rocks. Suddenly the whale was off the rocks and in the water. Halvdan took the whale a good bit out to where the water was deep and only there did he untie the rope and set the whale free. The villagers were holding their breath again as they watched the whale. For a while it remained motionless, but then it began swimming, away from the beach and into the open water where it rejoined the whales waiting in the fjord. Its return to the sea was applauded with yelling and laughter. The first whale was saved.

For hours, the villagers worked hard to keep the remaining whales cool while they carried out the rescue operation to get them all off the beach. They worked slowly, each time making sure that the ropes were well tied around the whale before Halvdan was given the signal to start backing up the boat. The whole time they also kept an eye on the whales that had been rescued and were now swimming in the fjord, anxious that they should not beach themselves again.

Refloating the bigger animals first turned out to be a good strategy because all the rescued whales joined the whales out in the fjord and didn't swim back. The whole operation took a long time and the sun disappeared behind the mountains while the villagers were still working, but there was still enough light and most of the onlookers stayed at the beach. Around 10:00 p.m., the last whale was pulled out and a feeling of euphoria spread among the people following the spectacle.

"We were incredibly relieved," says Sigrund, looking at some of the pictures spread out in front of her on the table. "People were yelling 'bravo' and 'hooray' as we saw the last whales swimming out to the waiting herd. It was an experience you never forget."

Why the killer whales beached themselves that day remains a mystery. Strandings of live whales are rare, and strandings of killer whales even more rare. Most whale strandings involve sperm whales, pilot whales, or a close relative of the killer whale, the pseudorca. These are all species that normally live in deep oceanic waters and only very seldom come close to the coast. One theory is that since their sonar is attuned to the wide-open ocean, these more oceanic species can become confused in shallow waters, which can lead to "mistakes" and eventually a stranding. Another theory is that strandings are caused by the strong ties between individuals in these highly social species, and that mass strandings occur when an animal deliberately beaches itself, maybe to avoid drowning because it is sick and weak, and then the whole group follows it. But even if strandings at least on the surface look very much the same, it is probably a mistake to search for only one explanation. Traffic accidents can also look very much alike, but there can be many underlying causes that are widely different from each other. Lack of attention, exhaustion, a child demanding attention in the back seat, the state of the road, the behavior of other cars, and mechanical failures can all lead to an accident. Likewise, strandings mostly likely happen due to many different reasons.

In the rare occasion when killer whales strand, usually the animals die. As with other cetacean species, there is no good explanation for the majority of the killer whale strandings. The exception of course is the strandings that sometimes happen in the groups of killer whales that intentionally beach themselves to catch seals or sea lions resting on the shore. This obviously involves a much higher risk of getting stuck on a beach, but it still happens, only very rarely. In the groups of killer whales that catch part of their prey in this way, it is often a few individuals from the group that specialize in this behavior. Some get very good at it, but it remains highly risky and can always go wrong.

One early morning in July 2015, a group of yachters sailing in Hartley Bay on the coast of British Columbia in Canada discovered a stranded killer whale lying on some rocks. The whale came from a pod that hunted other marine mammals. Killer whales in this area do not deliberately beach themselves when they hunt, but maybe the stranded individual had been swimming through an area with underwater rocks and had overestimated the depth of the water. When it was discovered, it was lying helplessly on the rocks, a long way from the water's edge. It was a female, and like the killer whales that stranded in Austnesfjorden in Norway, this killer whale was also calling out loudly for her group.

When she was found, her skin was already drying out and the people from the sailboats and other volunteers joining in the rescue organized a water pump from one of the boats so they could keep her wet and cool. The volunteers also covered her with a tarp and wet blankets so the sun wouldn't damage her skin. For more than eight hours, they worked tirelessly to keep her cool and moist while they waited for the tide to return and lift her off the rocks again. Around four in the afternoon the water was finally so high that it covered her and she started to move a little bit. Soon after, she succeeded in wriggling away and swam straight for her group, which had been waiting at a distance all day.

The whales in Austnesfjorden had also reunited with their family, which had been waiting for them during the many long hours of the rescue operation. In the days after the stranding and the successful rescue, the local villagers saw killer whales in the fjord, in the area where Kjartan Krane had first spotted them the day of the stranding. But they stayed out in the middle of the fjord and didn't come close to the shore again. For Sigrund, the day and the experience with the killer whales changed her view of the species, which in the years before the stranding had been the object of intense hunting in Norway. It became a species she felt something different for, something special.

"Some years after that episode, I was with my family on vacation in southern Europe and we went to an aquarium where they had both dolphins and killer whales," she says, looking up from the kitchen table. Her eyes meet mine, and she looks both sad and angry. "It was terrible to keep them in such a place. They shouldn't be there at all," she says. "The experience with the stranded whales did that to me. I couldn't watch it."

The selfless act of helping the killer whales back into the sea and the joy it brought to the people who took part in the endeavor raises some interesting questions. Why do we bother helping animals at all? What in the behavior of animals—or maybe just in their being—elicits our compassion enough that we go through the trouble of picking up a bird that has fallen out of its nest, bottle-feeding a prickly hedgehog whose mother has been hit by a car, or raising a fawn in our living room despite the obvious inconveniences? The answer seems to be that we recognize ourselves in others. Even though whales live in an environment that is completely foreign to us, and it is hard for us to imagine what their life is like, we still connect to them.

Humans are empathic beings by nature, and despite the species barrier, we identify with helplessness, despair, fear, and other emotions in animals—especially mammals, which we are evolutionarily closest to. Empathy brings out compassion and a desire to help. For social animals like humans, it makes sense in an evolutionary perspective that empathy within the tribal group begets compassion and action, as it ensures the survival of our children and keeps our societies together. It does not, however, stop with our own children or even our own species. Every day you can read stories in the media of people helping animals. This wish to help vulnerable animals has led to the foundation of wildlife rescue centers, but even in the absence of such facilities, I am guessing that most of us, just like the people in Austnesfjorden, would help or at least try to help if we found a beached whale or another animal that it was in our power to help.

This is not a new observation. Even Charles Darwin recognized our inclination to extend sympathy and help beyond our own species: "*Sympathy beyond the confines of man, that is, humanity to the lower animals, seems to be one of the latest moral acquisitions. . . . This virtue, one of the noblest with which man is endowed, seems to arise incidentally from our sympathies becoming more tender and more widely diffused, until they are extended to all sentient beings*," he wrote in his book *The Descent of Man*.[2] A recent study concludes that empathy and compassion between humans and other species may have evolved because it offered us evolutionary advantages.[3] The authors argue that our ability to recognize another species' needs, plus our inclination to help or rescue animals who are suffering or in danger, may have roots in our own biology. Further, the evolution of interspecific compassion may have contributed to our own survival. Perhaps the capability to understand other creatures made our ancestors better hunters, or they became better at judging the dangers wild animals presented. Helping wild animals could have been the first step taken toward the domestication of species that we now take for granted as pets or farm animals.

Animals help animals from other species, too. There are examples of the adoption of an orphan by a mother from a different species, even between species that are not closely related, like marmosets and capuchin monkeys, lions and leopards, and bottlenose dolphins and melon-headed whales. There are also examples of animals helping other animals out of dangerous situations. One of the most interesting involves humpback whales intervening in killer whale attacks on other whale species, on seals and sea lions, and even on a sun fish.[4]

The interactions that we witnessed between these two species while doing fieldwork in Norway were always relatively peaceful, at least on the surface. Perhaps the humpbacks knew that the Norwegian fish-eating killer whales presented no threat to them and that they were big enough to chase them away if needed. But in other places, where killer whales regularly eat other marine mammals, humpback whales and, especially, their calves are vulnerable to killer whale attacks. Often the adult humpbacks will defend the calves by escorting them and positioning themselves between the calves and the killer whales. This, of course, makes sense from an evolutionary perspective as they are protecting their own kin, but surprisingly humpback whales do not stop with their own calves or even their own species.

In one case, a humpback whale interceded in an attack on a seal by a group of killer whales in Antarctica. Two marine mammal scientists, Robert Pitman and John W. Durban, witnessed the incident in which the killer whales cooperated to make a wave to flush the seal off an ice floe:

> *At one point, the predators succeeded in washing the seal off the floe. Exposed to lethal attack in the open water, the seal swam frantically toward the humpbacks, seeming to seek shelter, perhaps not even aware that they were living animals. (We have known fur seals in the North Pacific to use our vessel as a refuge against attacking killer whales.) Just as the seal got to the closest humpback, the huge animal rolled over on its back—and the 400-pound seal was swept up onto the humpback's chest between its massive flippers. Then, as the killer whales moved in closer, the humpback arched its chest, lifting the seal out of the water. The water rushing off that safe platform started to wash the seal back into the sea, but then the humpback gave the seal a gentle nudge with its flipper,*

back to the middle of its chest. Moments later the seal scrambled off and swam to the safety of a nearby ice floe.[5]

The authors of the study suggest that the behavior may have developed as a sort of "spillover behavior." Originally the humpback whales' behavior would have benefited their own kin, but if there are few risks or costs to the behavior, there is no selection pressure preventing it from "spilling over" to another species.

Perhaps the ability of our own species to feel empathy and compassion is also the result of such a spillover effect? The success of our own species depends on our ability to take care of our young, and this urge is so strong that we sometimes extend it to other species. To the villagers in Austnesfjorden, it was the realization that the beached whales were just like us—adults taking care of their young—that moved them to save the killer whales.

CHAPTER 8

The Whales at the End of the World

In 1741, Commander Vitus Bering sailed forth from Avacha Bay in Kamchatka, Russia's most easterly province, to find new lands. Until then, eastern Russia was considered the edge of the known world. In 2019, after flying from Copenhagen to Moscow and bumping from airport to airport in Siberia for a total of 33 hours, I arrive at this distant outpost in the middle of the afternoon. I land in Petropavlovsk-Kamchatsky, and it does feel very much like the end of the Earth. It isn't, though; it's merely the first stop on a journey that will take me to the southeastern part of Kamchatka to meet some of the most isolated and dedicated killer whale researchers in the world.

Petropavlovsk in Avacha Bay is a peaceful place to start a journey. The bay offers a shield from the hammering waves and furious winds of the North Pacific. Commander Bering must have thought the same when he picked the bay as a departure point for his prestigious expedition to conquer new lands and win fame and fortune for the Russian rulers,

Tsar Peter and Empress Anna. Indeed, it was so protected that when Bering, after almost eight years of preparation, was finally ready to hoist the sails, he had to wait five days for enough wind to fill them and get out of the bay. The expedition consisted of two ships, both custom-built for the voyage, in the small settlement that Bering founded in the northern part of the bay. He called the settlement Petropavlovsk after the ships, which he named *St. Peter* and *St. Paul.*

Once the expedition had left Avacha Bay, they had no shortage of wind. In the inhospitable seas between Kamchatka and Alaska, violent storms battered the ships. With fog to contend with as well, the two ships lost sight of each other just two weeks after their departure. They would never be reunited. The *St. Paul,* captained by Aleksei Chirikov, limped back to Petropavlovsk five months later. With sails in tatters and sailors dying from scurvy and exhaustion, they made it back just before the arrival of winter.

Bering himself was not so lucky. On October 4 the expedition's naturalist, Georg Steller, noted in his journal, "*We were as well accustomed to storms as we were to daily deaths.*"[1] All but a few, Steller among them, were so sick with scurvy from lack of fresh food and water, that they were unable to perform their duties on board and resigned themselves to meet their deaths in their bunks. The ship itself was barely able to move—the sails were shredded to pieces and the men on board too sick to sail it. Steller logged another blow just a few days later. On October 6, they were out of brandy. After six months of continuous battle with the North Pacific, the *St. Peter* was wrecked on a small island in early November.

Hoping that they had come back to the coast of Kamchatka and unable to save the ship, the surviving crew members scrambled ashore and watched as their vessel was battered to pieces in front of them. They soon learned that they were stranded on a hitherto uncharted island. Stoically they dug holes in the sand dunes to shelter in and prepared to overwinter on the island, which is now called Bering Island.

Bering had to be carried ashore, too sick and weak to move on his own. Despite efforts to revive him with freshly caught game and a tiny offering of edible vegetation, he kept getting weaker. On December 19 (or December 8 by the pre-Gregorian calendar), almost a month after they had landed on the small island, he drew his last breath. The remaining officers and crew buried him along with the others, who one by one succumbed to scurvy, exhaustion, and other ailments.

Meanwhile, Steller was doing quite well, under the circumstances. In the midst of his shipmates' misery, he continued to hunt, bringing fresh meat back to the desolate camp at the beach and encouraging everybody to eat that, rather than the moldy provisions salvaged from the ship. He was convinced that scurvy could be kept at bay with fresh meat (we now know that he was right).

On my way to the harbor in Petropavlovsk, I visit a monument honoring Bering for his voyages and the discovery of the open passage between Russia and Alaska, today named the Bering Strait. Hidden in a corner of a public park, the monument doesn't attract a lot of attention. It is a tall column, maybe shaped to look like a ship's mast with a globe balancing on the top. A plaque at the base simply gives the dates of Bering's birth and death, 1680–1741. More prominently placed, in an open place near the harbor and the bay, is a large monument with the apostles St. Paul and St. Peter holding a huge orthodox double cross along with a gigantic statue of Lenin overlooking the bay. There are no monuments celebrating Steller's contribution to science or to the survival of the remains of Bering's expedition.

A local man named Alexander meets me at the harbor to take me by boat some hours south of Petropavlovsk where the killer whale researchers have made their camp. We communicate as well as we can using Google Translate, but when we head out of the bay into the open sea, Alexander must concentrate on steering through the massive swells. As we proceed south, we follow a jagged coastline; tall cliffs in vivid shades of cream, ochre, amber, coal grey, and black drop sheer to the sea where they are pounded endlessly by the surf. In the background, a number of volcanoes reach up to the skies. With their impeccably formed cones crowned with snow, they look almost too perfect, like children's drawings of mountains.

The richness of marine life must have been exceptional in Steller's time. We are so used to living in a world that has been systematically hunted, trapped, fished, burned, logged, and harvested that we have little idea how abundant the wildlife was before so many of us humans came along. Kamchatka still evokes a sense of this prehistoric richness. No doubt there is massive fishing in Russian waters, but here on the surface at least, the abundance of wildlife is impressive. Myriads of sea birds fly by continuously and other myriads bob on the surface water and plop under when we approach. There are guillemots, horned puffins, kittiwakes, shearwaters, cormorants, fulmars, skuas, and eagles. A couple of humpback whales surface as we cruise by. I'm not surprised that we pass no signs of civilization as we head south. The area can be viewed on Google Maps, but there is no "street view" to speak of; there are no roads, no towns, no harbors, and no settlements even along the entire coast.

I have been invited to visit the camp run by FEROP, the Far East Russian Orca Project. Strictly speaking I have invited myself, but Tatiana Ivkovich, who runs the camp, has generously agreed to put me up. After a couple of hours in the big waves, Alexander rounds a green headland sticking out into the sea and points to the top of it. I see the tents high up on a plateau, scattered in the green grass. They look small. Miniscule, in fact. As we come into the small bay next to the cape, people come out of the tents and start climbing down the steep path to the shore. They carry out a small inflatable that had been lying on the shore and they row out to Alexander's boat to pick me up. Moments later, he waves good-bye and leaves with a roar from the outboard engine.

We climb back up to the plateau and settle down in the biggest tent for a cup of tea and introductions. Tatiana, or Tanya as she is called in the Russian tradition for diminutives of names, is a petite woman with bright blue eyes and blonde hair in a long braid down her back. Despite her slight figure there is an air of authority about her. Besides Tanya, there are three other women. Asya is in charge of the acoustic project, collecting data on different killer whale groups' dialects. Sasha is a new student who will be looking at the conservation status of the Kamchatka killer whales, but this summer she's acting as Asya's field assistant. Julia is a visitor like me, a friend of Tanya's who has come to spend time in the camp and see killer whales. The three men around the table all work as volunteers on the project. Dima and Vanya play chess; Zhenia is following their moves. They all look very focused on the game.

Tanya has been doing research in the Avacha Gulf for almost 20 years. She studied in St. Petersburg and still lives there with her husband and two daughters when she is not in Kamchatka. She came to eastern Russia for the first time in 2002 and has been back every summer since except for two years when her children were too small—a situation I recognize and can relate to.

"In the beginning, we were on a small island further north, but then we moved the whole camp to this headland at the mouth of the fjord," she says. "We like it here. It's protected from the prevailing winds—also from bears—and there is fresh water." They call it Zelyony Mys, the Green Cape, she tells me. It's an apt name.

"This tent is our kitchen and our workroom at the same time," she explains. I look around at the shelves brimming with canned food and bulging cardboard boxes, with vegetables hanging in nets from the aluminum poles supporting the shelves. Life jackets hang from the poles in the roof along with more nets full of cabbages and potatoes. On the table are a few computers and a bowl of dried fruit and nuts. She asks if I want something to eat and when I admit that I am quite hungry she points to a down jacket thrown on a chair. "There's a pot under the jacket, to keep it warm," she explains. "Just grab a bowl and help yourself."

After the meal of beans, meat, and potatoes, she shows me around the camp. A hose brings fresh water from the stream, which also serves as a fridge for the meat I've brought as a gift. In a plastic container it will keep cool for a few days. In addition to the tent where we had our tea, there is a tent for sleeping and a tent for washing, which they call the *banya*. The sleeping tent accommodates six, in very narrow bunk beds, so I am quite happy when Tanya points out my own private sleeping quarters. It's a small igloo tent at the top of the slope overlooking the fjord to the west. It's not big, but there is plenty of room for me and my small backpack. Dog-tired, I go to bed before nine o'clock and only vaguely notice the wind and the rain during the night.

The next morning it is still raining and too windy to go out on the water. Instead, I sit down with Tanya to learn about the Russian killer whales and her involvement in the project. She tells me the project started with Erich Hoyt, a longtime conservationist from the Whale and Dolphin Conservation Society, and Haruko "Hal" Sato, a Japanese researcher, naturalist, and photographer. Together, Hoyt and Sato sought a third person to initiate research on killer whales in the Russian Far East. They found Alexander Burdin, a researcher from the Pacific Geographical Institute Far Eastern Branch in Petropavlovsk. On a scouting trip in the waters off the Kamchatkan coast in 1999, Burdin and Sato ran into killer whales aplenty. And a few years later they kicked off the first studies of Russian killer whales, founding the Far East Russian Orca Project.

When Tanya first heard about the project, she wrote to some of the Russian partners in Petropavlovsk asking if there was any way she could participate. They said yes, and at short notice, she cancelled a planned family vacation in Crimea to join the project.

"From the beginning, we focused on using the same methods that had worked so well in Canada and the US," Tanya says as she rises to fill the thermos with more tea. "We took ID pictures and tried to find out which whales were traveling together, who was related to who. What were they doing? What were they eating? Et cetera."

"Do you know how many killer whales there are in this area?" I ask.

"We have identified about eight hundred from this area, the Avacha Gulf, and the eastern Kamchatkan coast. There are other populations of killer whales in the Okhotsk Sea, in the Kyril Islands, and around the Commander Islands (Bering Island, Medney Island, and a number of smaller islands). Of the eight hundred, we know two to three hundred quite well; we recognize them when we see them and because we have followed them continuously for so long, we also know a lot about their social structure."

She takes a sip of tea and continues, "There are about twenty-five families. In many cases we know how the individuals in a family are related. Who are sisters or brothers. Who is their mother. Sometimes we even know who the grandmother of a particular whale is."

Tanya goes on to explain more about the ecology and behavior of the killer whales in Kamchatka. The social system is very similar to what is known from other areas, such as the Pacific Northwest, where killer whales have been studied for more than 40 years and the life history of many individuals is known in detail. I express my admiration for FEROP's work. Photo-identifying 800 killer whales is no small task, and the in-depth knowledge the team has of the killer whales' social organization is second only to the body of knowledge about the Pacific Northwest population.

I ask what the killer whales in Kamchatka eat.

"Most of them are fish-eaters," explains Tanya. "It used to be Atka mackerel (*Pleurogrammus monopterygius*), but they disappeared due to overfishing and now the whales have switched their diet to salmon."

I mention to her how Norwegian killer whales hunt herring cooperatively.

"We saw the same here," Tanya says. "They circled the schools of Atka mackerel and used tail slaps to kill them. But now when they eat salmon, they hunt individually."

"Do they have marine mammal–eaters here, too?" I ask.

"Yes," she confirms. "We have two ecotypes of killer whales here, both fish-eaters and marine mammal–eaters." The mammal-eaters' population turns out to be quite small, only about 200 individuals. The research team has used DNA to confirm that they are not just behaviorally different from the fish-eating killer whales but also genetically distinct. This is crucial information in regard to conservation of the species. It means that the mammal-eaters don't interbreed with the fish-eaters, and because they are so few, their status is more vulnerable.

"We know most of the whales in the area," Tanya says, "but occasionally we see an individual or even a family we haven't seen before. Just a week ago we had such an encounter." She stops herself. "Wait, I will show you some pictures."

She pulls out her computer and looks for images of a male they saw just a week before my visit.

"Here it is," she says as an image pops up on the screen. "It came into the other fjord you can see just over there," she says pointing to the southernmost of the two fjords visible just outside our tent to the west of Green Cape. "From our lookout up here we noticed the animal, but it was behaving very strangely. It was a lone male, and it's very rare to see killer whales alone. If you wait, its family will usually be around. But this one was clearly not in company of other whales. We were worried because it was not swimming or diving; it was just lying there."

She continues her story as she pulls up more images of the whale. "We put one of the rubber boats in the water and approached slowly, but as it was clear that something was wrong, we didn't go too close for fear of stressing it. As we came nearer, we could see that its dorsal fin was full of puncture wounds."

Tanya finds a close-up of the fin. I agree it looks very odd. From top to bottom there are rows of holes, all of them deep puncture wounds into the flesh. When she zooms in, I can see fluids running from the wounds, either blood or lymph fluid. The distances between the holes are evenly spaced. It looks like bite marks from an animal with big teeth spaced quite far apart.

"Another killer whale?" I ask astounded.

Tanya nods and tells me she is certain that the wounds have been inflicted by another killer whale or whales. "But we don't know why," she says. "Was it a stranger that was chased away? Was there something in its behavior that elicited aggressive behavior in others? Was it sick and weak before it happened and was that why it was attacked?"

She describes how the whale was lying almost motionless at the surface. "It was breathing with difficulty and was clearly unable to swim or dive for food, not moving; just being washed over by the waves every now and then."

The team stayed nearby the whole day and observed the male carefully. They feared that it might have suffered internal injuries as well as the wounds to the back fin, but realizing that there was nothing they could do, they kept their distance. As the day ended, they reluctantly went back to the camp.

"The next morning when we looked out, it had disappeared; we don't know if it died during the night or recovered enough to swim away. But if we see it again, we will have no problems recognizing it. With those wounds, there will be scars forever."

There are not many accounts of aggression between killer whales, but this may be more common than we think. Some of the many nicks and scars that can help researchers to identify one killer whale from another may very well come from intraspecies aggression—something well known in many other species—especially between males. Generally, males have bigger scars, nicks, and marks than females. Maybe they are just more prone to get markings because their fins are so much bigger, but it is also possible that some of the nicks are from males fighting other males.

The next day the weather has improved, and we get ready to head out to sea. After a breakfast of porridge and tea, we all help carry the waterproof cases of equipment down to the beach and load them into the boats before clambering on board. We head due east on perfectly calm water. My expectations are high, but they are not met immediately. We drift around, surrounded by hordes of seabirds, but there are no whales of any kind. After a while we lower a hydrophone in the water, but apart from the noise of distant ships, there are no underwater sounds either. We drink tea, eat biscuits, and wait. It is rather pleasant—the rubber boat gently rocking in the water and the bird life offering plenty of entertainment.

After a while Tanya gives up waiting and grabs the VHF radio. She speaks Russian with someone, and shortly afterward Dima starts the engine, and we begin to head south.

"I spoke to the boats you can see down there," she explains and points to a couple of big cabin cruisers on the horizon. "Those are tourist boats; they have killer whales with them," she explains.

When she sees my baffled expression, she laughs, "Yes, we have whale-watching here, too," she says. "Tourism is growing, and they've added whale-watching to their list the last few years."

So, with the whale watchers' help, we find the killer whales. They are some five to six miles south of the Green Cape, spread out over a huge area. They seem to be foraging. Every now and then we see one making quick jerks with its body, followed by a big splash. Tanya explains to me she wants to get to the feeding site as quickly as possible, so each time we see a splash, Dima speeds up and we hurry to the spot.

Two team members grab a small fishing net each and lean over the bow of the boat. The nets look like, and probably are, children's fishing nets for catching bugs and tadpoles in a pond. Here they are used to catch fish scales. Tiny specks of glimmer, scales from the salmon that have just been caught are visible in the water after a kill, but they sink quickly, so only the nimble-fingered has a chance. Despite my best efforts I don't catch any, but Tanya catches several almost every time. Sasha, who has been taking notes of where we are and what behaviors we are seeing, puts away her notebook and grabs a box of vials and a pair of tweezers. Meticulously, she plucks the small fish scales from the net and places them in a vial of alcohol, where they gently sink to the bottom.

"We send these for analysis by colleagues at the university," Sasha tells me. "They can tell us which species of salmon the killer whales are eating, both from the shape of the scales and also by extracting DNA from them."

The results have given the researchers a good understanding of the killer whales' feeding preferences. Interestingly, killer whale females and males in Kamchatka have different diets. The females eat almost exclusively coho salmon, while males eat a mix of about 50% chum salmon and 50% coho salmon. Occasionally males also take Chinook, pink, or sockeye salmon. Halibut is also eaten sometimes and there have been incidences where killer whales have taken these fish from longlines.

"In 2013 and 2014, times were bad," Tanya tells me. "There were very few salmon, and we saw fewer killer whales. Some families disappeared and have not been seen since. Some family members died."

This is disturbing and echoes the demise of the Southern Resident killer whales in Puget Sound on the other side of the Pacific. There, a small genetically isolated population of killer whales has been declining since 1995, dropping from almost 100 whales to only 73 in 2022. Like their conspecifics—or members of the same species—in Kamchatka, they rely on a diet of salmon, especially Chinook. Scientists suspect malnourishment as a cause for the decline. The Southern Resident killer whales are now on the endangered species list, but appearing on a list does not ensure their survival.

When we get back to the Green Cape, I take a walk on the small beach below the headland. The vegetation is lush and most of the species look familiar. Knotweed, monkshood, and meadowsweet grow on the slopes running down to the shore. Maybe they are not identical to the European species I know, but they must be closely related. Black guillemots and spectacularly colored harlequin ducks putter around in the shallow water.

To my surprise there are sea otters, too. It is the same species that lives on the North American west coast and they are as charismatic as their American cousins. Through my binoculars, I observe a female sea otter with a pup. The small pup whines pitifully when the mother leaves it alone at the surface and crawls quickly on top of her when she resurfaces. The sea otter is another species that Georg Steller described.

"*They embrace their young with an affection that is scarcely credible*," he wrote in his field notes.[2] He recognized them when he saw them on Bering Island, having seen them previously in Kamchatka, where they were called "sea beavers" by the Itelmen, the native people of Kamchatka. On Bering Island, sea otters became important for the survival of the marooned sailors. Appallingly easy to catch as they slept innocently on the rocks, Steller could walk right up to them with his ax and club them, serving not only the meat but also the entrails to the reluctant sailors. For months it seemed to be an inexhaustible resource. But of course, it wasn't, and a handful of hungry men was all it took for the local population to go into a decline. After a few months, the men had to go further and further to find them. Steller also noticed that the otters had become more watchful and therefore much more difficult to kill. It doesn't take long for animals to learn where danger comes from.

Steller hunted for more than fresh meat. He hungered for knowledge, too. Whenever he had had the opportunity during the voyage, he had hurried ashore to collect whatever he could find. Plants, rocks, or animals—all went into his bag to be carefully inspected and described once he was back on the ship. Even under the dreadful conditions on Bering Island, he continued to add to his collections. As a result, a whole menagerie of animals from the North Pacific region was first described by Steller and bear his name today: Steller's jay, Steller's eider, Steller's sea lion, Steller's eagle, and the now extinct Steller's sea cow, which he found on Bering Island itself. He also discovered new species of plants, fish, and mollusks. Most importantly for the stranded sailors, it was Steller's diligence that secured the return of the remaining crew. Under his guidance, they built a smaller ship from the wreckage of the *St. Peter*, and in 1742 the survivors of the expedition reached the safety of Avacha Bay again.

The sea otters around Green Cape are on their guard, keeping their distance from me on the beach and equally from people in boats. The many larga seals (a close relative of harbor seals) popping their heads up to follow my steps along the coast are also cautious.

"Good for you," I think. "It may not be safe to come too close to people here." Before departing for the Green Cape, I had seen shops in Petropavlovsk selling furs of lynx, saber, wolverine, and bears, so I imagine that seals and sea otters are also likely in danger.

Later, back in camp, I see Dima and Zhenia busy sawing driftwood. It looks as dry and hard as rock, but they manage to cut up a little pile. I ask what the wood is for.

"The stove in the banya!" they tell me.

My eyes grow bigger. "The stove?" I ask.

I draw the tent flap to the banya, and sure enough there is an old-fashioned cast-iron woodstove in there on which two big pots are heat-ing water to boiling point. It is almost as hot as a sauna inside, and after some instruction on how to mix the hot water with cold water using an array of buckets and pots, I later get a wonderful warm shower by emptying them over my head. The water just runs onto the ground. I contemplate what it reveals about the Russian soul that they allocate the woodstove to the banya instead of the unheated communal tent, where we sit with hats and down jackets on to keep warm. Something good, I think. To give so much priority to washing with hot water is a fundamentally good thing.

We go out to sea whenever the weather permits, which is not every day. Most days the killer whales are foraging and feeding, like the ones we encountered the first time. Often they are spread out over a large area, with each whale hunting by itself. Tanya recognizes the majority of the whales when we encounter them and knows what family they belong to. One day, we come across a group of eight whales, four adult males, two adult females, and two juveniles. Each time we witness a successful salmon kill, we hurry to get the scale samples. At the end of the day Tanya declares that we have collected samples from all the individuals in the family except a female with a characteristic notch in her dorsal fin. She is determined to get a sample from this female as well, but it takes us a few more hours before we finally do.

The light is therefore low when we finally finish and head back toward the Green Cape. The outboard engines on the inflatables are small and it's a long journey home. On the way back, I ponder how vast the North Pacific is and how quickly bad weather can blow in. I have looked in vain for safety precautions like a spare outboard engine or a GPS navigator. At least the weather is calm as we chug slowly along. Then, still about an hour's drive from the cape, we encounter a fleet of low speedboats heading south at very high speed in the fading light. They come in pairs or trios, maybe 9 or 10 boats in total. I realize that they must have come all the way from Petropavlovsk, as there are no other harbors in the area. Dima and Tanya exchange a few words in Russian, and then we adjust our course off to the side. The tension is palpable.

"Who are they?" I ask Tanya.

"Poachers," she replies. "They are not nice people; we don't want to cross their path."

Killer whales are not the only ones hunting for salmon. Poaching is a colossal and increasing problem in Kamchatka. When the Soviet Union collapsed in 1990, unemployment and poverty hit hard, harder in Kamchatka than elsewhere because of its remoteness and lack of infrastructure. Many took up illegal fishing just to survive, and there were plenty of salmon at that time. But poaching was small-scale then compared to the industry it has turned into. Now it is big business. It is organized. Ruthless. And completely unsustainable.

The object is not so much the fish itself but the hidden treasure inside it: the salmon's roe, called *krasnoye zoloto* in Russian, meaning "red gold." The eggs are perfect spheres, glistening orange and large; the bigger they are, the better the price. Poaching for the roe is like a gold rush, no less feverish than the search for the golden mineral. The value of a salmon's roe is 10 times that of its flesh, and it's a lot easier to handle than a whole fish. Salmon poaching has become the main income of many Petropavlovsk residents and has made some rich beyond their dreams. As an environmentalist says in a short documentary on the poaching: "*When you see a Hummer parked outside a chopper pilot's house, you know right away that he bought it with money made transporting salmon caviar.*"[3]

Most poachers only take the roe and dump the fish. If the poachers are operating in the open ocean like the ones we encountered, they dump the fish in the sea. If they're doing their business in the rivers where the salmon run, they dump the fish on land in big heaps, where they rot and attract bears. The roe just needs salting and shipping to be ready for the dinner tables of Moscow, Korea, Japan, and China. It is estimated that poaching accounts for as many salmon as the legal fishery, which is itself considered unsustainable, taking as it does about 100,000 tons of salmon each year. In any case, it severely depletes the stocks of all five species of salmon: Chinook, chum, coho, pink, and sockeye. Roe that ends up in restaurants and on dinner tables will never grow up to become an adult salmon, returning after years of maturing in the Pacific to the river where it originated to complete the cycle. It is a dead end. The consequences can be dire for the whole ecosystem, where the salmon is a cornerstone species.

Until the Russians started the war in Ukraine in February 2022, tourism was growing in Kamchatka. Wilderness and volcanoes, spiced up with the highest density of brown bears in the world, were the main attractants, but killer whales could easily have become equally important for the tourism industry. In the Pacific Northwest, where whale-watching has been well established for decades, it is estimated that it generates more than $100 million a year. The ripple effect, including what visitors spend on transportation, hotels, shopping, and so on, amounts to maybe five times this figure. Another effect is the desire of both visitors and the tourism industry to protect the valuable objects of interest people come to see: the killer whales, the bears, the salmon, and nature in general. The potential for whale-watching in Kamchatka is huge; there are many whales, they are present in the summertime, and the landscapes and wildlife are nothing short of amazing. With the international unrest and boycott of Russia, this potential development has collapsed.

Before this, there was already concern that another kind of catastrophe is around the corner—for the salmon, the bears, and the killer whales. Tanya and her team saw signs of trouble in 2013 and 2014 when there were too few salmon and the killer whales disappeared. If the salmon stock is severely depleted, it may affect the killer whale population badly.

But this is not the only bad news. There is more.

This chapter is dedicated to the journalist and environmentalist Masha Netrebenko. Her brave filming of the whale jail in Srednyaya Bay catapulted the captive whale industry in Russia into the awareness of big media and global consciousness. She died in 2020 after a snowmobile accident while covering a dog sled race in Kamchatka.

CHAPTER 9

The Whale Jail

On November 6, 2018, a disturbing video appeared on Facebook. Filmed from a drone, the video showed a bird's eye view of a series of ten enclosures, or sea pens, in the water of Srednyaya Bay along the easternmost Russian coast, south of Kamchatka and close to Vladivostok and the Chinese border. In the video, three big floating shacks can also be seen next to the open pens. Even from the height of the drone, it is possible to make out what is in the sea pens. They look like white tadpoles in a crowded petri dish, but in fact they are beluga whales—up to ten in each of the pens. The cramped size of the pens and the number of animals in each leave little room for the whales to move around. They swim around each other in circles. The commentary to the video reveals that there are also 12 killer whales and five walruses confined in sea pens inside the three floating shacks.

The video was filmed and uploaded by Russian journalist Masha Netrebenko, who at the time was based in Petropavlovsk where she produced television for Vesti Primorye, the regional branch of the Russian State TV and radio. She agreed to tell me the story of her visit to Srednyaya Bay and how she managed to get the drone footage. When we connect on Facetime, I'm looking at a young woman with shoulder-length dark-brown hair wearing a hooded sweatshirt and black-rimmed glasses. She greets me with a big smile and perfect English. She tells me that before her visit to Srednyaya Bay she had heard rumors about the captive whales there and had decided to check it out. She didn't know what to expect or even where exactly to find the facility. But once she arrived in the windswept landscape, there were few buildings to choose from.

"The whole compound was surrounded by tall fences that went down to the shore and into the water," she tells me. "That particular day the tide was so low that it was possible to go around the fence and get into the area where the whales were kept."

Masha says that she was tempted to take advantage of the low tide and the opportunity to document what was inside the facility, but after weighing the risks in her head, she decided to stay outside. She knew the guards were heavily armed.

"Instead, I climbed a small nearby hill from where I could overlook the compound. I decided to set up a drone and film from above."

It didn't take long before the guards at the enclosure spotted Masha and came out to confront her. But despite the men's intimidations, she continued to film. She knew that she was not guilty of trespassing and hoped that the guards would not provoke a violent escalation. Forced to fly the drone high because it lost signal every time she flew it lower and closer to the sea pens, Masha soon realized, too, that the facility was jamming communication to the drone. But it wasn't necessary to get close to see what was going on. When Masha had heard the rumors about the captive whales in Srednyaya Bay, she had not imagined that there were so many.

"When I was done, I hurried away and I uploaded [the footage] the same evening," she says. Originally she wanted to produce a longer piece for television, but when she saw the video, she changed her mind and uploaded it to Facebook right away. She suspected it had the potential to spread on social media.

She was right. The response on social media was immediate. The video was picked up by national press agencies in Russia, and in a matter of days it swept across big international media like CNN, CBS News, Forbes, and National Geographic. Viewed by millions, the shocking images of the whales in crowded sea pens sparked outrage, and the place was quickly dubbed "The Whale Jail" and tagged #freerussianwhales, #letthewhalesgo, and #freeOrcasAndBelugas.

The reports of the live capture of Russian killer whales was one of my reasons for coming to Kamchatka and visiting Tanya's research group at the Green Cape. They know the Russian killer whales better than anyone, and I was keen to hear their views and learn more. One evening after a long day at sea, I sat down with Tanya and asked her to tell me about the live capture of killer whales in Russia. She gives me a funny look.

"Where do you want me to start?" she asks. She smiles briefly, but her eyes are not smiling. She looks tired but obliges and continues. "It's a long story and it goes back a long time. Do you remember the group of whales we saw this afternoon?"

I nod. That afternoon we had had a great encounter with at least 25 whales moving north while they were catching the occasional salmon unfortunate enough to cross their path. The whales were spread out across a huge area and, as on most of the days we were at sea, we tried to photo-ID as many as possible and get samples of fish scales after a kill. We took photos of a mother with a little calf. The calf was playful and was often swimming at the mother's side, maybe trying to nurse, but the mother seemed more interested in catching salmon.

"Well, that particular family has a history of live capture. A bad one," Tanya says.

She stands up to get a bag with a couple of external hard drives for her computer. After a bit of searching, she finds the right one and opens the file she is looking for. It's a video from an inspector on a Russian whale capture ship. I move to her side and we look at it together. On the screen I can see that it is a foggy day with calm waters, not unlike our day at sea. The video shows what is believed to be the first live capture of killer whales in Russian waters. It took place on September 26, 2003. A long line of buoys is floating in the water next to the ship. It looks like a purse seine. Purse seine nets are often set with dinghies going out from ships and are used to encircle schools of fish, such as herring, tuna, or anchovies. Once the perimeter of the net surrounds a whole school of fish, fishermen cinch the bottom of the net and the ship hauls the catch on board. Purse seine fisheries often have a large amount of bycatch because they haul everything that is inside the net.

However, the net in the video was not set to catch fish. The aim was a much bigger target and the people on the ship have done well. Inside the perimeter of the net, which is maybe 75 m across, swims a group of killer whales. It looks as if there are about 25 whales, a couple of big males, many females, and several calves. There is at least one very small calf. The whales are swimming in tight formation at high speed, and when they get to the net and realize that it's a barrier, they change direction abruptly. Their exhalations are loud and sound strained. They appear distressed and anxious. A couple of whales are slapping their tails at the surface, often a sign of frustration and anger.

Tanya stops the video and points to a big male in the group.

"That whale is Hookey, a whale we know very well. He is one of the first whales we photo-IDed here in Kamchatka." She points out his characteristic fin and markings on the saddle patch. "It was his family we saw today in the Avacha Gulf. He is dead now, but his family is still here. They were all in the net. We saw two of his sisters today. We also saw Prizrak today; she was the female with the little calf. She is also here in the net on the video."

Tanya explains that there are at least three different families in the net. All families they know well from the waters in the Avacha Gulf.

We continue watching the video. A couple of whales try to break out of the net. They swim toward the rim at high speed, and I hold my breath because it looks like they are going to escape by weighing down the floats and swimming over the net. But they don't. Instead of escaping, they become entangled. A female struggles desperately to free herself, whipping the water to a foam in her panic. Then she tries to dive under the net, but when she resurfaces, she has become even more entangled; the net is twisted around her fin. At one point she hurls herself at the net, twisting her body around as if she is trying to swim through it with all her power. But the net is too big and too stretchy for her to push through, it merely bulges with her efforts. She pauses as if to gather her strength before making another futile attempt to break out.

A smaller whale, a calf or a juvenile, perhaps hers, is close by. In the video a man is shouting in Russian. Tanya explains to me that he is yelling "*Ne uydut, Ne uydut!*" ("They are escaping!"), and then "*Zastryal, zastryal!*" ("They are stuck!").

When another man replies, Tanya translates his words as something along the lines of, "So what. It drowns. Fuck it."

Eventually the female in the net stops moving. She lies still at the surface with the net draped around her dorsal fin. She has either given up or her strength has run out. Then the video cuts to a new scene of the whales on the other side of the net. I look at Tanya. She shakes her head.

"She didn't make it. She drowned."

The fate of the other whale that also became entangled in this catch is unknown. Officially only one animal died during the capture.

A few whales hang about outside the net. Some may have escaped the net, others are probably family members that dodged the purse seine when it was set and are now standing by, unable to do anything for their relatives inside the net but also unwilling to leave.

On the ship a winch starts up, and the men on board pull in the net. Two small whales try to get out by vigorously slapping their flukes, but they are slowly and inevitably dragged aboard. The next video clip shows a whale hung in a sling on board the ship. The catchers hose it down with sea water.

Tanya tells me that the people on the ship took one whale, a young female, and released the rest of the group, including Hookey and Prizrak. The young female they kept was later transported to the aquarium in Utrish on the coast of the Black Sea. She died there 13 days later.

Russia is the only country involved in capturing and trading killer whales today. They stepped into the commercial vacuum that presented itself when other countries dropped out of the business. The first killer whales in captivity came from the Pacific Northwest waters of the United States in the 1960s. In 1970, an operation in Puget Sound went catastrophically wrong and five animals died. Public outrage forced legislation to put a stop to the live capture of killer whales in US and Canadian waters. This didn't stop live capture though; it merely pushed it to other parts of the world.

In Iceland a 6 m long killer whale was caught by accident in a herring purse seine net in 1974. Too big to handle and too old to tame, the captors released it when they couldn't find a buyer for it. Animal trainers and traders prefer young animals that are easy to train and small enough to be transported both on land and by air. But the incident fueled interest in live capture as a business in Iceland. One year later, two animals were caught, and the year after that four. Some 10 years later, a total of 59 killer whales had been caught in Icelandic waters. Some were released and some died, but most made it to a final destination. No matter where in the world that destination was, they all ended up in a tank with chlorinated water. Many of the killer whales that can be seen today in oceanariums and amusement parks in North America and Europe are, or descend from, Icelandic whales.

When Icelandic captors ceased their business in 1989, it moved on to Russia. Despite the first failure in 2003, live capture has continued in Kamchatkan waters and in the Okhotsk Sea between Kamchatka and mainland Russia since. And the operators learned a lesson in 2003. They stopped allowing observers on board, and the information about how many animals have been captured and what happened to them is sketchy: One animal was captured in 2010. At least seven killer whales were captured in 2012 and 2013. Another seven or eight whales were caught between 2014 and 2017. The whereabouts of these whales are not known in detail. Some of them are in Russian aquaria and amusement parks, some in Chinese. Some, like the whale caught in 2010, are reported as escaped from captivity. But the math doesn't add up. More whales have been caught than are registered in aquaria in Russia and China. "Escape" in many cases probably means that the animal died while in a holding pen or during transport.

The year 2018 started out promising for Russian whale captors. Despite protests from environmental groups and scientists, permits were issued from the local branch of the Federal Service for Supervision of Natural Resources to capture 13 killer whales and 90 belugas. The captors were a conglomerate of four different companies working together. As the summer of 2018 drew close, they had everything ready for the grand slam of a total of 103 whales. They had the permits in their pockets. They had onshore camps on the coast of the Okhotsk Sea with wooden towers erected to look out for whales and fast speedboats equipped with nets to surround them. They had a big ship with cranes to lift the whales out of the water and containers on board in which to keep them. They had the holding facility down south in Srednyaya Bay with all the sea pens. They even had guns.

They only got one thing wrong—and boy, did they get that one really wrong. They miscalculated the public aversion to their business. When the environmental group Ocean Friends in Sakhalin learned about the planned captures of 2018, they launched an expedition to document the process and raise awareness about live capture of marine mammals. Exasperated with the lack of governmental control with the captures (there was none), Ocean Friends established what they called "public environmental control." Their goal was to film and document as much as they could about the whole operation and to communicate it to a wide audience.

A small group of seven dedicated volunteers went to the Okhotsk Sea where they set up camp at the coast. From there they went out in small boats with outboard engines to find the captors and film what was going on. On the coast they found deserted camps the captors had used in previous years, with iron containers built for the transport of animals. The containers were small, not much more than 3 m to 5 m long. Evidently the captors were targeting juveniles and calves.

Eventually Ocean Friends found the operation's ship in the Okhotsk Sea, but the men on board refused to answer questions or let them come near their ship. When the Ocean Friends volunteers set up a drone to film the ship from above, the crew on the boat shot at it. Ocean Friends intended to obstruct the captures, but they had to give up because their boats were too small and too slow to catch up with the captors' speedboats. And besides, after what happened to the drone, they knew that the captors were willing to use their guns.

When the Ocean Friends activists returned to their camp at the coast, they found that it had been ransacked. Everything had been thrown on the ground, the gasoline for their outboard engines had been poured out, and what wasn't destroyed was stolen. The Russian Far East is a wild place in more ways than one, but at least it meant that the captors perceived Ocean Friends as a threat. Maybe they were beginning to understand that publicity could be a hazard to their business.

In the summer of 2018, the captors succeeded in catching all or nearly all the animals they had permits for. If any animals died in the process, which most experts believe to have been inevitable, it was not reported and probably just led to the captors having to catch another animal. Each successfully captured whale was lifted with a crane onto the ship and taken to the western part of the Okhotsk Sea. Here they were loaded into containers on trucks and driven more than 1,000 km south to Srednyaya Bay, where the holding pens were ready for them.

For a while, things were quiet. The four companies that had captured the whales had filed for export permits and presumably already had customers waiting for the animals in China. But their plans were beginning to unravel even before the whales reached the holding pens in Srednyaya Bay.

Environmental organizations in Russia continued to fight the sluggishness of the authorities. In November 2018, Ocean Friends, who documented the capture in the Okhotsk Sea, together with two other environmental organizations in Sakhalin, Boomerang and Sakhalin Environmental Watch, filed a lawsuit against three different Russian government agencies, including the Russian Federal Research Institute of Fisheries and Oceanography (VNIRO), for negligence in relation to both the capture and the treatment of the animals after the capture.

Prompted by media attention and the activities of the environmental groups, the prosecutor's office in Sakhalin opened an investigation into the legality of the captures. And there were plenty of issues to consider. Russian law prohibits the capture of marine mammals unless it's for scientific or educational purposes and then only for use in Russian aquaria, but local authorities in the far east of Russia had chosen to disregard the law and grant permits to the captors anyway. The trade in marine mammals with another country is banned. The captors treated this detail as merely a formality. Despite having already filed for export permits, they declared that the animals were to be displayed in Russia only—in their own facilities—and provided fake programs to demonstrate how they intended to use the animals. But it was evident that the Russian facilities didn't have enough space for the over 100 whales.

The Pacific branch of VNIRO and the governmental organization overseeing these declarations, tactfully turned a blind eye to these rather blatant fabrications and went ahead with a recommendation of the catch anyway. VNIRO had played a dubious role a few years earlier when their former director of the Pacific branch, Lev Bocharov, was involved in the illegal capture of two killer whales and five belugas, which were subsequently sold to aquaria in China. He was found guilty by the Sakhalin Court but apparently the system reacted lethargically to these activities and still allowed things to slip through their fingers.

Another important issue was the type of killer whales the captors had in their holding pens. Thanks to the efforts of researchers like Tanya and her colleagues, it had been established that there are different populations of killer whales in Kamchatkan waters, and some of these populations are vulnerable or endangered. The marine mammal–eating ecotypes are much less numerous than the fish-eating ecotypes, and in the spring of 2018 the marine mammal–eating killer whales were placed on the Red List (formally the International Union for Conservation of Nature's Red List of Threatened Species) in Kamchatka. This means that it is illegal to catch them. The population may be as small as 200 individuals. The captors were aware of this but didn't care.

The prosecutors' investigation wanted an answer to these questions, but to get those answers they needed experts to confirm what kind of whales were caged in Srednyaya Bay. Tanya was one such expert, and in November 2018 she was summoned to inspect the facility in Srednyaya

Bay. Before the visit she was required to sign a confidentiality agreement, and she is still not allowed to discuss or share her observations from that visit or even tell who arranged it.

A few months later, in January 2019, she was asked to come again, and this time with Olga Filatova, a research fellow with the faculty of biology at Moscow State University and also a part of the FEROP's project. Several veterinarians were also summoned. This time the visit was a more open process.

So as not to put Tanya and her colleagues in jeopardy, I only discuss the second trip with her. The first question I ask her is who initiated the second visit.

"It is not a secret," she says. "We were invited by Dmitrij Lisitsyn from Sakhalin Ecowatch, who were involved with the whole case that was unfolding at that time." She and the other visitors flew to Vladivostok and drove in cars from there. We look at a map of the area. She traces the route with a finger and describes how isolated the place was. It is an area with few settlements and long and lonely roads.

"It was bitterly cold when we arrived. The fences and the knowledge that the guards were armed made it scary, and I was relieved to see a few familiar faces among the other people who had also been asked to come."

They were met by a large group of armed men who led them inside the facility. Since November, when Masha Netrebenko's footage emerged showing the pens in the blue waters of the Srednyaya Bay, temperatures had dropped drastically, and during Tanya's visit ice held the facility in a strong grip. The water in the outside pens with the belugas was partly covered with ice, making it difficult for the belugas to find enough space to surface and breathe. Staff had to work ceaselessly with metal bars to chop away the ice, which formed almost as quickly as it was broken up.

The killer whales were inside the three shacks under roofs, giving them a little bit of protection from the fierce weather outside. Tanya recalls the sad sight inside. More than twice as big as beluga whales, killer whales need a lot of space to swim normally, but inside the shacks, they were squeezed together in three small enclosures. Metal bridges around the enclosures allowed the staff and visitors to walk close to the surface of the water in the enclosures. Of the original 12 whales caught, there were now only 11. One had "escaped." And even though the shacks gave some protection from the harsh weather, the whales were swimming in what could best be described as icy slush.

"The killer whales became agitated when we entered," Tanya remembers. "They swam to the side of the pens and poked their heads up, begging for food and contact." Tanya knows that spy-hopping to many people looks endearing and cute, but she emphasizes that here it was a sign that the highly social and intelligent whales were utterly bored. They had nothing else to do. Tanya and Olga set to work with the other invited experts.

There was no doubt in their minds that the killer whales in the enclosures all belonged to the red-listed marine mammal–eating ecotypes. To an expert eye, there are small differences in the form of the fins and shape of the body that immediately reveal the ecotype. To document this, Tanya and Olga photographed all the individuals in detail and took DNA samples to establish their affinity genetically. They also took ID pictures of all the animals, important work both to establish if the whales were known individuals already in an ID catalog and to be able to track the whales' whereabouts if they should once again become free.

ID pictures are also an important tool to establish which whales go where, if they end up in China or in an aquarium in Russia. The researchers also measured the length of all individuals to get an estimate of their age. They were all small, juveniles or calves, but one animal—a female named Alexandra—was especially small. It was very likely that she was less than one year old when she was caught (in which case it would have been illegal to catch her).

The veterinarians took microbiological samples and microchipped all the individuals. Some of the whales did not look in good health; their skin was peeling off in big flakes and they had round lesions stemming from either bacterial or fungal infections or frostbite. Most afflicted was a young male named Kirill. Not only did his skin come off in big chunks, but he was also inactive and lay motionless at the surface most of the time. His breathing was shallow and slow. The researchers worried that he had pneumonia and needed immediate treatment.

Tanya and Olga also monitored the behavior of the whales and recorded their underwater vocalizations with a hydrophone, but most of the time the whales were actually vocalizing above water. Tanya explained that this was different to her first visit. During the intervening months, the whales had probably become increasingly oriented toward humans and were therefore vocalizing more in the air. Tanya was especially concerned about Alexandra.

"She was calling out all the time and another whale in one of the other enclosures would often respond to her calls. I saw her try to get into the other whale's sea pen by chewing on the net. It was difficult to watch and heartbreaking to see such a small whale separated from her mother and so desperate for contact."

After two days of data collection and monitoring, the researchers left the facility. I asked Tanya how the visit affected her. She hesitated, searching for the right word.

"It made me feel shame," she finally concluded. "When I saw the animals in there trying to get out . . . it shouldn't be like that." She paused and then added, "Killer whales in the wild are so powerful and independent. To see them there, trapped and begging for food . . . it was just wrong. I felt ashamed."

Outside Srednyaya Bay the pressure on the captors and the authorities who backed their operations was building, both nationally and internationally. The drone footage that Masha Netrebenko had uploaded started a landslide of protests. From Sakhalin in the easternmost part of

Russia to Moscow in the west, people signed petitions and participated in public demonstrations. Millions signed petitions in the rest of the world, too. International superstars like Leonardo DiCaprio, Pamela Anderson (of *Bay Watch* fame), Jean-Michel Cousteau (the son of the famous ocean explorer Jacques Cousteau), Jane Goodall, Bill Branson, and numerous others protested against the "whale jail," advocated for the release of the whales on Twitter, and wrote to Russian president Vladimir Putin directly asking him to intervene. The court case raised by the environmental organizations' lawsuit in November 2018 was already unfolding. In February 2019, another lawsuit was filed, this time by Russian officials from the Border Guards Department. The prosecutor's office in Sakhalin, who surveyed the facility in Srednyaya Bay, charged the companies involved in the capture and captivity of violating animal welfare laws, of illegal capture of the endangered ecotype, and of catching underage animals.

You may be forgiven for suspecting that eastern Russia is often a lawless, free-for-all society where corruption and big business wins. But in the case of the "whale jail," you would be wrong. The court's ruling, which came in May 2019, was a triumph for the environmental organizations who had raised the issue. It was also a victory for their lawyer, Nataly Litsitsyna, who led the case, and an enormous encouragement to the millions in Russia and abroad who had signed petitions and protested against the capture.

The judge ruled that the capture of the whales was illegal. The killer whales were of the marine mammal–eating ecotype, which were on the Red List. The captures took place in an area where captures were not permitted, and the captors had fiddled with the ships' logs to cover this up. The captures and transport of the animals had taken place without the oversight of a veterinarian. Five of the beluga whales were judged to be underage and still depended on their mother's milk. As a result of the court's ruling, all the captive animals were confiscated. The decision to release them back to Okhotsk Sea, from where they had been captured, came shortly after.

The four different companies involved in the capture operation and in the Srednyaya Bay facility were given heavy fines, totaling more than 150 million rubles, or 2.5 million US dollars. In 2021, the ban on capture of killer whales was extended to 2023. It is unclear if the four companies would return to business as usual should they manage to get a permit.

It is a sad truth that nothing has contributed more to changing attitudes toward killer whales than having killer whales in captivity. Showing killer whales in aquaria and oceanaria swung the pendulum from hate to love, from fear to admiration, from being seen as a menace to being seen as icons of beauty and wilderness. It wasn't until the first killer whales in captivity proved to be both intelligent and trainable that they also became loveable. The earliest captive killer whales were accidental catches and usually didn't survive for long, but some of them lived long enough to enthrall both their trainers and the people who came to see them—and long enough to establish the industry, one that didn't wait for killer whales to be accidentally caught, but that actively hunted whales for aquaria and amusement parks.

It is a paradox that when people watch killer whales in oceanariums or amusement parks, what they fall in love with is not a true reflection of a killer whale. Few of the behaviors that you see in a show are seen in the wild, and most of the behaviors that characterize killer whales in the wild will never be seen in captivity. This, you may argue, is true for all animals in captivity. But it is extraordinarily true of killer whales. Sometimes I wonder if that is precisely what is attractive about zoos and aquaria: wild animals bereft of anything that makes them wild. Do we like nature better when it is toothless and benign rather than when it is wild and untamed?

Our attitude about what is acceptable to put in a cage at a zoo has not been constant. In the late nineteenth century, some zoos exhibited not only the usual menagerie of big cats, crocodiles, elephants, camels, brown bears, and hippos but also "exotic" humans. The Copenhagen Zoo, not too far away from where I live, as late as 1909 displayed Inuits, Chinese, Indians, and Africans, complete with their traditional clothing and other ethnographic attributes. The idea of displaying humans in a zoo is, of course, outrageous today. In a few decades from now, maybe we will shudder in disbelief that we ever kept our nearest relatives—chimpanzees, gorillas, and orangutans—behind bars in zoos. And maybe there will be a time in the not-too-distant future when the idea of keeping whales and dolphins in a zoo seems equally outrageous.

In many western countries, we are certainly moving toward an understanding that some species should not be kept in zoos. Killer whales are one such species; they are too big, require too much space, and have social structures and behavior too complex to justify having them in captivity. In the aftermath of the television documentary *Blackfish*, about a killer whale at SeaWorld that killed three trainers, SeaWorld has been under immense pressure to give up displaying killer whales. In 2016, they declared that they would end their killer whale breeding programs. Given that they are no longer involved in live capture, they are in effect phasing out SeaWorld's big black-and-white killer whale trademark.

But there are still killer whales in many other establishments in both Europe and North America, and not all of them are following in the footsteps of SeaWorld. The growing trend in public concern about the welfare of captive killer whales has not yet impacted China, for instance. They have more than 80 marine parks and about 40 more under way. Marine mammals are the pinnacle of attraction there and killer whales more than any other species.

Even in the wilderness that is the Green Cape, bits and pieces of news trickle down to Tanya and her team through messages on the satellite phone or through visitors who, like me, arrive in the middle of the field season. Therefore, she is up-to-date about the release of the first killer whales from Srednyaya Bay. At the time of my visit, seven of the ten killer whales had been transported back to the Okhotsk Sea and released. But three still languished in the whale jail. Reports in various media outlets confirmed that the releases went relatively smoothly and the killer whales seemingly did well. However, it was cause for concern when a video appeared showing the little female, Alexandra, approaching fishermen and begging for food. Was she not able to find her own food? Had she become so accustomed to human company that she preferred it to her own species? Thankfully, some weeks after her release she was spotted swimming with three older killer whales also released from the whale jail.

At least one of the released killer whales joined a group of wild whales. A female named Vasileyevna was observed swimming with a small group of wild killer whales approximately a month after her release and was even seen participating in a seal hunt and sharing the food with the other whales in the group. So a return to a wild life is possible even after more than a year in captivity.

My visit to the Green Cape culminates with an extravagant sunset framing the black cone of the Viluchinskiy volcano at the end of the fjord against a sky flaming in red and purple. We stand outside the tent in awe of the scenery. It seems like an appropriate finale to a visit of a place that is truly wild. The next day, Alexander—who drove me out to the Green Cape in his little boat—comes back to pick me up, and we bounce our way back to Petropavlovsk.

Evidently, the explorer Ferdinand Magellan didn't sail these northern waters when he named the new sea he encountered in 1520 the Pacific Ocean. The water splashes across the deck as we hammer the waves with a force that seems sufficient to pulverize the little boat; between that and the big waves, I suddenly catch a glimpse of a small group of killer whales surfing in the breakers. It looks like they enjoy the rough weather more than I do.

Back in the civilization of Petropavlovsk, I enjoy the comforts of a real bed in a small hotel with the free Wi-Fi that connects me to the rest of the world. I look for an update on the last three killer whales from Srednyaya Bay still in captivity. It turns out that while I was at the Green Cape, they too were released. They were loaded into trucks that took them down the long road back to the Okhotsk Sea where they were set free near the place where they were captured. The number of released killer whales then amounts to 10, which is one short of the original 11 killer whales that were in captivity when Tanya visited Srednyaya Bay in January. The eleventh was reported to have "escaped" a few weeks after Tanya's visit, though the captors didn't give any detailed information about which individual was missing.

In contrast to that report, staff from the prosecutor's office were allowed to go inside the enclosures about a week after Tanya's visit, and they glimpsed a killer whale floating on its back, belly up, in one of the pens. They filmed it and showed it to the scientists waiting outside, who were banned from entering. The scientists are convinced that the animal was Kirill, the young male that looked so lethargic and sickly on Tanya's visit just 10 days earlier.

In June 2019, before the release of the whales, Vladimir Putin appeared on a national television phone-in where the public were invited to ask questions. In a few sentences, he boiled down the problem with the live capture industry: "*The killer whales alone—as far as I know—are worth around $100 million*," Putin said. "*When there is big money at stake, problems are always hard to solve.*"

Later in the fall of 2019, 90 beluga whales were also released, representing a lost income of millions of dollars for not just their captors but also the aquaria they were destined for. If managed cleverly, however, they also represent incredible value and potentially an even bigger income as wild animals. Businessmen and women need only to cross the Bering Strait and visit the thriving wildlife tourism in North America. Increasingly, people in Russia are no longer willing to accept that their wild animals are caught and exported to other countries. There is a growing understanding about the value of wilderness, not just in terms of what it means for tourism but also the value it has for people in Russia.

The potential for conflicts is huge, and the potential for income drives not just politics but also criminal enterprises, such as the salmon poachers I saw in the Kamchatkan waters and the illegal killer whale captors. At least this time, big money wasn't enough. Maybe it is a hopeful sign that the influence of the skilled and energetic environmental groups in Russia is mounting. With Russia isolated internationally because of the war in Ukraine at the time of this writing, the environmental groups are forced to continue their work in isolation as well. Their proficiencies and determination will become more important than ever. The demand for live killer whales to the entertainment industry in Asia has not disappeared.

CHAPTER 10

When the Hunters Become the Hunted

A friend of mine grew up in a small settlement in Greenland with a Danish mother and a Greenlandic Inuit father. Surrounded by the tundra, the family hunted grouse and reindeer, and in the nearby creek, they could pull up as many trout as they wished. When he was still a little boy, his family moved to Denmark. Thousands of kilometers away from his childhood home, everything was vastly different and foreign—the windy weather, the houses, the flat landscape, the traffic, and the people. But the thing that bewildered him the most was that nobody hunted the many fat pigeons that he saw near the central station in Copenhagen. The birds only moved half-heartedly when someone approached them and would clearly be easy to catch, he thought, and they looked delicious—much better than the liver paste he was learning was a main staple of the Danes' diet.

This story illustrates a key difference between the Danish and Greenlandic ways of life. The Danish think of nature as a backdrop for their outdoor activities and food as something they buy in a supermarket. To Greenlandic people, on the other hand, nature and food are intimately linked. The Kingdom of Denmark colonized Greenland and its Inuit people in the late 1700s. Bound together by the centuries-long history of colonialism, the two peoples are so widely different in their ways of life that, at times, it can be difficult to find common ground. This is increasingly true as many Greenlandic people today desire greater autonomy or complete independence from Denmark.

Most Danes have ties to Greenland. Either they had an uncle who worked there for a couple of years, or a schoolmate who was from Greenland, or a cousin who married a Greenlandic man or woman. Often these Danes' connections to Greenland are characterized by a deep love of the country and its people. But the love is not always reciprocated. Many in Greenland feel that stereotypes concerning them and their way of life run deep in Denmark, simultaneously romanticizing the iconic Inuit, with their polar bear pants and kayak, while patronizing them. People in Greenland have seen the Danes getting all the good jobs, higher wages, and influence in powerful positions, which has led to growing resentment of the dominance of Danes in their country. Although the Greenland Home Rule went into force in 1979 and the Self-Government Act in 2009, Greenlanders do not have complete autonomy.

Like many other Danes, I also have a relationship with Greenland, having traveled and worked there many times; sometimes visiting family who've lived there, other times doing research or participating in scientific or educational projects on whales. In the late autumn of 2021, I traveled to Tasiilaq on the eastern coast of Greenland, a wilderness so wide and isolated that there are only two small towns along the 2,500 km coastline. Tasiilaq is the southernmost town. Ittoqqortoormiit, the other small town, is 800 km to the north of Tasiilaq.

Originally, I had arranged to go there with Rune Dietz and Christian Sonne, professors at University of Aarhus in Denmark. They study the contaminant levels in different species of marine mammals and had planned to collect samples in Tasiilaq in the summertime, when the hunting of whales and seals is the most active. But the coronavirus pandemic and canceled flights messed up those plans, and I ended up going there on my own.

People hunt killer whales on both the eastern and western coasts of Greenland, and my objective was to learn more about the Inuit hunt of killer whales on the eastern coast. The intensity of the hunt fluctuates; some years, no killer whales are taken, other years as many as 44 have been killed.[1] In some places, like Tasiilaq, local hunters consider the whales a pain in the neck with their reputation of scaring away important game animals, like seals, narwhals, and belugas. However, in Tasiilaq, the hunting of killer whales has increased over the last two decades, and biologists at the Greenland Institute of Natural Resources have warned that it is uncertain whether these catches are sustainable.[2] Scientists don't know how many killer whales inhabit the waters outside Tasiilaq, and they argue that killer whale populations are often relatively small, therefore any hunt could potentially be dangerous to the population's status. Another concern is that killer whales in other places, like the Pacific Northwest where they are closely followed by researchers, occasionally join each other in "super pods," wherein most, if not all, of the individuals congregate in a big gathering. If the same held true for the east Greenlandic killer whales around Tasiilaq, an unlucky strike of coincidences could potentially wipe out a major part of this population. Losing a population could mean more than just losing their role in the ecosystem, or losing that genetic diversity; it may also mean losing that population's culture. All over the world, killer whales have been characterized by having cultures, which are the behaviors they learn from each other that are distinct from others of their species.[3] If all the members of a social group with a unique culture disappear, so does that culture.

Tasiilaq looks as if a child has decorated a model mountain with toy houses in primary colors. The roads are too steep, the grass too green, the houses too colorful, and the mountains too pointy. Many houses are flanked by a couple or a trio of snowmobiles as well as wooden dog sleighs, and often the family's huskies are chained outside the house. Sometimes the dogs have a little shelter they can go into if the weather is harsh, but most often they remain in the open year-round. The neighboring house to where I stay has a dog with puppies, and they dart out to meet me whenever I pass by, wagging their tails and licking my hands. I quickly found out that the dogs are usually fed in the late afternoon. As soon as someone starts to feed their dogs, all the other dogs across the entire town pick up the activity and howl in synchrony as if they all belong to the same pack.

But not all is idyllic. On the way to the harbor, I pass a garbage bag lying at the roadside, the innards spilling out like the entrails of a dead animal. Empty bottles and cans festoon the curbs, and social problems are everywhere. Outside a supermarket, I notice a group of people who are obviously drunk. Upon talking to people in town, I learn that many are deeply concerned about the common incidences of alcohol-related troubles, the high rates of unemployment, and the lack of opportunities for young people. Just before my arrival, tax returns and pay day had put extra money in people's pockets. With the extra money, the sales of alcohol increased, as did the number of domestic violence cases. During the same time period, a woman was raped, two young women committed suicide, and another four attempted suicide. To halt the calamities, local authorities closed the sale of alcohol in all shops, bars, and restaurants for a two-week period. In a small community with only about 2,500 people, such tragedies affect almost everybody. One of the women who committed suicide was a maid in the small hostel where I am staying; the other worked for a woman I met later.

Surrounding the town with a raw wilderness, the Arctic tundra embraces Tasiilaq from all sides. From the northern rim of the town, I follow a footpath leading into the wilds through the Valley of Flowers. I don't know if the valley is named after the Arctic flowers that bloom there in the spring and summertime—but not in September while I'm there—or if it is named after the eternal flowers, of which there are thousands, in the churchyard that stretches along a creek in the bottom of the valley. In the simple burial ground, row after row of plain wooden crosses line up humbly in the yellowing grass, but the graves are demonstratively decorated with plastic flowers in screaming red, orange, pink, and yellow, as if the bright colors can defy the reality of loss and grief.

The name of the valley and the graveyard is also the title of a novel by Niviaq Korneliussen that was awarded the prestigious Nordic Council Literature Prize in November 2021, a few weeks after my visit there. Central to the novel is the story of a young woman's suicide in Tasiilaq. Korneliussen writes about the epidemic of suicides among young people in Greenland, where the rate of suicide is twice as high as in any other country in the world,[4] and the lack of will among everybody—from schoolteachers, family, and social workers, to health authorities and local governments, all the way up to the parliament and the national government—to do anything about it, or event talk about it. In an interview with a Danish newspaper, Korneliussen argues that the high rate of suicides in Greenland is related to Greenland's history of colonization and the loss of culture and identity that followed in the wake of the transformation of Greenlandic society in the 1950s.[5]

Denmark colonized Greenland in a gradual process. During the eighteenth century, numerous trade stations and small settlements were established along the western coast of Greenland, with Danish station managers governing each little settlement. Initially, the trade stations were quite respectful of the Inuit traditions—principally because the items they were trading, like fur, sealskins, walrus, and narwhal tusks, came from the Inuit—so the Danes felt it was important not to change their traditional way of life too much. Later, especially in the twentieth century, Danish governing of Greenland shifted, and a massive transition took place. Many small hunting settlements shuttered and the people moved—or were moved—to bigger towns with schools, industrial facilities, apartment blocks, and a modern way of living. This meant hunting was swapped for office jobs or employment in various industries, and a natural economy exchanged with a monetary one.

Only 100 years ago, practically all adult men in Greenland were hunters and their families depended on the fish and meat they brought home. Today only around 7% of the men are full-time or professional hunters, but many continue to be leisure hunters. To most Greenlandic people, hunting is still at the heart of what defines them as a people, and the transition to a more Western way of living comes with a tremendous cost. Social workers and scientists believe that the intoxicated people I saw outside the shop, the domestic violence, and the graves of the two women who committed suicide are directly related to loss of identity that comes with losing the traditional way of life.

Tobias Ignatiussen is one of those professional or full-time hunters in Tasiilaq. The professionals have a state-issued license to hunt both mammals and birds and to sell the meat to others, something that is often done in an open-air market called the *brættet*, or "the board," referring to the coarse cutting boards where the meat is presented and cut for the customers. The hunters are, of course, obligated to follow the hunting regulations, report their catches, and for some species, comply with quotas. The professional hunters get their main income from hunting and fishing, but many of them supplement this with other activities. In cooperation with a partner, Tobias has started a nature-based adventure company taking tourists and other visitors to the waters outside his hometown. He has agreed to let me accompany him and instructed me to meet him in the small harbor in Tasiilaq.

The harbor is the town square. There is an almost constant activity at the filling station's two gasoline and diesel pumps as men are preparing their boats to go fishing or hunting. A little group of men and women are sitting on a bench sharing a couple of cigarettes. Cars are driving by, unloading people or just stopping to chat. Like an extension of their arms, rifles are invariably carried in one hand, sometimes a thermos or a plastic bag in the other, as men board their boats. Not all the men go alone either. Some of them bring their family, including small children and grandparents. A family outing here almost invariably is a trip out at sea.

At the harbor, a large container ship docked on the northern pier dominates the scene. I am told that it is the last delivery ship before winter, and it is bringing all the supplies for the supermarket, the hardware shop, and the few other shops in town. All other boats in the harbor are small dinghies that are tied to a couple of floating docks. They look tiny and utterly unsuitable for the open Arctic Ocean.

Tobias greets me on the dock, and I climb into the little boat. It's a small fiberglass dinghy with a sprayhood. Under the sprayhood is a broad shelf with extra clothes, some blankets, and a few plastic bags. I set my camera on the shelf as Tobias puts the boat in gear, and we set off. It's a windless day with sunshine from a clear blue sky; still the air is cold, and I quickly dive back under the sprayhood to find extra clothes.

I don't speak any Greenlandic, but Tobias speaks Danish well enough that we can communicate. He tells me that we are heading to an area east of Tasiilaq where there is a good chance of encountering whales. As we leave the protection of the bay, I expect that the open waters will be windy, but here, too, it is perfectly calm. Out in the open there are only the flat blue ocean and icebergs that have separated from any number of the many glaciers in the area. A haze erases the horizon, obliterating the point where the sky and the sea fuses in a powdery mist. The icebergs are spread out in front of us, tall and white, like high-rise buildings in a city of blue.

Tobias drives the boat as if he knows which iceberg the whales are hidden behind. On the flat water, we move quickly, passing one iceberg after another. About 15 miles offshore we encounter the first whales. Tobias points them out; they are fin whales. They appear to be feeding, coming to the surface regularly to breathe, then diving under again for several minutes but staying in the same area. Growing up to 25 m long, fin whales are the second largest animals in the world, surpassed only by blue whales. With their long and slender bodies, they have been labeled the "greyhounds of the sea," as they can swim incredibly fast.

The fin whales are swimming in groups of two to four individuals, and I estimate that there are at least 15 animals around us. Tobias tells me that he expected to find humpback whales, too, but for some reason they are evading us. As they have in many other places, the humpbacks have become more common here over the last 10 years, and they can now be seen throughout the summertime in the waters outside Tasiilaq. But they are not the only whales that have become regular visitors in the area. Toothed whales, such as pilot whales, killer whales, white-sided dolphins, and white-beaked dolphins are seen here throughout the summer.

"These species were never here when I was a kid," Tobias tells me. It did happen every now and then, he explains, that someone encountered a group of killer whales, but it was extremely rare. Tobias believes that the occurrence of the newcomers is a result of climate change, and he reckons that they are most likely following different species of fish, like mackerel and herring, that are spreading north. To people in Greenland, global warming is a reality that is already affecting their lives profoundly. The ice cover in the fjords comes later in the winter and it doesn't stay as long as it used to. This means that the traditional hunting in the fjords, which is done from dog sleds and is crucial for catching seals resting on the ice, is now only possible in a brief window of time that gets shorter and shorter year by year.

Over thousands of years, the access to eastern Greenland has been hindered by *Storisen* (the Great Ice), a massive belt of sea ice stretching from the North Pole to the southern tip of Greenland. Formed in the Polar Sea, the ice is carried by the north polar current southward, effectively blocking access to the Greenlandic coast except for a few short summer months where the ice is thinned enough to allow the passage of ships. But like the ice in the fjords, the sea ice also disappears as temperatures rise. The warmer waters and ice-free conditions pull species like mackerel and even tuna into Greenlandic waters.[6] In their wake follows the killer whales, the white-sided dolphins, and the pilot whales.

When I ask Tobias how often he goes out to sea he laughs.

"Every day," he replies.

I understand that my question verged on stupid. To him the sea is not a place for leisure activities but the essence of his life. He hunts almost everything that is out here: eider ducks, various species of loons, geese and auks, seagulls, harp seals, ringed seals, bearded seals, porpoises, narwhals and pilot whales, killer whales, dolphins, and minke whales. Even the occasional walrus or polar bear. And he fishes, too, of course. The bigger whales, like fin whales and humpback whales, are plentiful but not hunted. To catch them requires a big boat with a grenade harpoon and there aren't any such boats in Tasiilaq or anywhere else on the east coast of Greenland.

I understand from Tobias that the hunting of pilot whales, killer whales, and dolphins is not so much a question of a deliberate decision as it is just a consequence of these species starting to appear in the area. When they arrived, they became game just like other wild animals on land or in water outside Tasiilaq. Since 2000, the Greenland Institute of Natural Resources estimates that around 40 killer whales have been killed locally. The number is based on the hunters own reporting and is likely much higher. Animals that sink are seldom reported.

When I ask what happens to the whales they manage to recover, he replies they are mostly used for dog food. However, a hunter north of Tasiilaq in Ittoqqortoormiit who fed his dogs killer whale meat reported that it was not good for the dogs' digestion. They became sick.[7]

Another reason to hunt killer whales is to limit their numbers in order to protect other species, like seals and narwhals, that the hunters are convinced the killer whales either eat or scare away. It doesn't help matters that narwhals have become scarce in eastern Greenland. Or at least the researchers at the Greenland Institute of Natural Resources think so, while the hunters' organization disagrees, claiming that there are plenty of narwhals, and if there is a problem, it is most likely due to the presence of killer whales, not the Inuit hunting.

In July, before my visit, Tobias had been further north of Tasiilaq in Kangerdlussuaq with a small group of hunters where they encountered a group of about 9–10 killer whales. He and the other hunters managed to shoot four of them, but they sank before they could drag them to shore. Tobias wanted to cut them up and look at their stomachs, as he was convinced that the killer whales had been eating narwhals and he wanted to document it. Another problem appears to be when the narwhals are scared by killer whales and flee into the pack ice or swim close to the glaciers in the fjords where the hunters can't follow them.[8] Some of the east Greenlandic hunters use traditional methods and hunt narwhals from a kayak, which can easily be destroyed by pack ice.

Believing that killer whales take much more prey each year than the hunters can catch, Tobias is convinced that the solution to the dwindling number of narwhals is to bring down the number of killer whales.[9] The researchers at the Greenland Institute of Natural Resources find no data to support the hunters' claims.[10] The killer whales are rarely in the same waters as the narwhals, and the stomach contents of hunted killer whales have shown them to eat seals, fish, and a few times also minke whales, but not narwhals.

People used to eat killer whale meat, and especially the *mattak*, a thin layer of skin with blubber, which is appreciated as a delicacy. But in 2018, the Greenlandic health authorities issued a warning to people in Greenland not to eat the meat from killer whales at all. They were adamant when they addressed the issue of eating the meat while pregnant, especially: "*If you wish to have children, you must take this warning seriously as the strongest effect of the contaminants will be felt in the next generations*" was their austere advice.[11]

The contaminants that the health authorities worry about are persistent organic pollutants like PCB, DDT, and a whole range of related chemical compounds, as well as the heavy metal mercury. It is not surprising that the level of these substances is high in the killer whales caught outside Tasiilaq. In fact, it is high in most killer whale populations and especially those who have a diet of other marine mammals. Rune Dietz and Christian Sonne, the two Danish professors whose expedition to Tasiilaq I had originally planned to join, have done extensive research in this topic and their results are deeply worrying. Before my journey to Tasiilaq, I went to see them in a white barrack on the premises of Aarhus University. Rune has dark hair with streaks of silver and a pair of glasses balancing on the tip of his nose. Christian is blond and burly. Like farmers and fishermen who have worked outdoors all their lives, they are both weather-beaten and one wears a fleece jacket while the other wears a woolen sweater. They look more suited to being in the field in the Arctic rather than in their university offices.

Rune digs out a reprint of a scientific paper from the prestigious journal *Science* and spreads it out on the table in front of us. Pointing at a series of graphs on one of the pages, he explains that he and the other researchers in the study grouped different killer whale populations according to the level of contaminants in their tissue. The tissue samples have come from different sources. Some were taken with biopsy darts like the ones I assisted Audun Rikardsen in collecting in Norway, some were from dead animals washed ashore, and others from animals that had been hunted in Tasiilaq and elsewhere in Greenland. The level of contaminants varies greatly. Least affected are the fish-eating killer whales in Antarctica, Norway, Alaska, the Faroe Islands, and Iceland. Worst are the killer whales in Gibraltar, the United Kingdom, Japan, and Brazil as well as the marine mammal–eating killer whales in the Pacific Northwest. Somewhat in the middle are the Greenlandic killer whales, along with those from the Canary Islands and Hawaii.

Rune explains that the killer whales' diet matters a great deal. Those that eat other marine mammals generally have 10 to 20 times the amount of contaminants in their tissue than fish-eating killer whales because the contaminants accumulated in the fatty tissues of fish, seals, and whales biomagnifies for each step in the long marine food chains. Bending over the papers, Christian, who is a wildlife veterinarian, points to the graphs of the different populations of killer whales.

"Knowing how PCBs impair reproduction and disrupt the endocrine and immune systems, we have calculated how different populations with different levels of PCB-burdens will develop over the next 100 years," he explains. "Unfortunately, the result is quite grim."

Their study concludes that more than half of the world's killer whale populations are at risk of collapse due to the long-term effects of contaminants, including the killer whales from Greenland, the United Kingdom, Gibraltar, and the marine mammal–eating killer whales in the Pacific Northwest.[12]

Between the fin whales surfacing, I discuss the hunt of killer whales with Tobias. He tells me hunting them can be tricky, and there has been an episode where a killer whale attacked a boat with a hunter.

"He survived," Tobias says. "You should get the story from him. His name is Gedion."

The following day I walk to Gedion's home, which my hostel's host told me it was the grey house next to the fire station. A female eider duck with feathers in an intricate pattern of brown and bronze lies peacefully on top of a plastic bag at the doorstep as if the bag is a nest and she is guarding a handful of eggs. Looking closer, I realize that the duck is dead, and it is not watching anything anymore. The duck will probably make a filling dinner later. Gedion Ignatiussen, like Tobias, is a professional hunter and brings home most of the meat his family eats. When I knock on the door his wife opens and after a bit of talking and gesturing to overcome the language barrier, she leads me inside to Gedion, who is sitting in the living room playing with a little boy. He agrees to tell me the story of the killer whale hunt that almost killed him, and we decide I should come back again the next day with a translator.

When I return the next day with the interpreter, Gedion welcomes us both inside to a coffee table and offers us cakes, soft drinks, and snacks. As we settle into the couches, I ask if Ignatiussen is a common name in Tasiilaq as Tobias's last name is also Ignatiussen. Smiling, Gedion explains that Tobias is his cousin, and that there are quite a few with that name in town. Then we move on to hunting and the question of killer whales in eastern Greenland.

Yes, he agrees, hunting killer whales has become quite common around Tasiilaq. In fact, just the day before, a friend of his had called him to ask for his help in a hunt. They got two animals. He scrolls on his iPhone and finds the pictures of them. Two killer whales, either young ones or small females, lie next to each other in a harbor with a small pier. Their striking black-and-white coloration is dotted with small splotches of red blood.

"It is in Kulusuk, an hour's drive from here," he explains.

He doesn't know what will happen to the whales but presumes that the meat will be given to dogs. Both the interpreter, a young woman with a newborn son at home, and Gedion were aware of the health risks involved in consuming the meat. Neither of them eats it, nor does Gedion's wife.

Gedion doesn't remember how many killer whales he has shot or killed; in any case, it is often a communal hunt where many men shoot at the same whale from different boats, so it is not easy to keep track. But he does remember the incident with the killer whale that Tobias told me about.

"That happened in 2015," he says, shaking his head in disbelief.

He tells me that there were five or six boats participating in the hunt and they had been chasing a group of killer whales for a while. A single boat can't handle such a large catch. Plus, killer whales move fast, so several boats have a better chance of success if they work together. When they closed in on a particularly large male, they had already shot two other whales from that group. Several men were shooting at the big male. Gedion, who was alone on his boat, fired his rifle, too, but the whale dove and disappeared below the surface of the water. When it came up for air, they fired again. This time the male did not recede, instead it accelerated and in a shocking assault rammed one of the boats with all its power, almost tipping it over. Then the big male sunk back into the dark water. The hunters were horrified, they had never seen a whale retaliate. Knowing their boats were small and the killer whale far outweighed them, they backed off, scared that the whale might try to ram one of the boats again.

"I went to the railing of my dinghy to look for it," Gedion says. "I didn't see him but suddenly he shot out of the water like a giant rocket."

He looks up as he continues the story.

"The whale spiraled itself out of the water," he says, lifting his arm in a swirling motion before tipping it over and smashing it down on the coffee table with a bang. "He landed right on my boat!"

The boat rocked violently and was dangerously near capsizing when the whale crashed down on it. The impact smashed the cockpit and hurled Gedion across the deck and into the railing on the other side. Lying on the deck, he looked on in disbelief as the big whale slowly slid back into the water. The other hunters fired a few more rounds, at last killing the whale, and Gedion limped back into the harbor with a wrecked boat, damaged hip, and a lot of bruises.

Still, there is admiration in his voice when he talks about the killer whales: They are smart—and beautiful too, he admits, but he believes their presence is problematic because they scare other animals away that are important to the local hunters in Tasiilaq.

There are places in Greenland where killer whales are welcomed. One place is on the other side of Greenland, on the western coast, and far to the north. Here at 77°30' N lies Qaanaaq, one of the northernmost towns in the world. Locked in ice most of the year, Qaanaaq is situated on a gradual slope facing the Inglefield Gulf, that stretches almost 70 km from its opening in Baffin Bay to its end in a large bay where several glaciers continuously dump colossal pieces of ice into the water. In the winter, the massive icebergs in the Inglefield Gulf sit immobile in the frozen sea, but as summer approaches, the sea ice breaks up and the icebergs regain their freedom of movement.

With the waterways open, the narwhals arrive, sometimes in the hundreds, other times in the thousands. They are small whales and, unlike their Arctic whale-cousins the belugas, their skin is not white but dapple grey, like those special horse breeds with mottled skin. Their tail flukes are characteristically heart-shaped, but most distinctive is, of course, the long spiraled tusk carried by the adult males.

Like wolves following bison, killer whales follow the trail of the narwhals. Contrary to the hunters in Tasiilaq, the hunters in Qaanaaq salute the arrival of the killer whales because they make it easier for them to catch the narwhals. Aleqatsiaq Peary, who is a hunter from Qaanaaq, was interviewed by researchers at the Greenland Institute of Natural Resources about his attitude to killer whales. He told them that killer whales help the hunters because narwhals are so afraid of them they begin to swim very close to the coast and the beach, where they are much easier to catch. "*Killer whales will neither be shot at nor killed in Qaanaaq*," Aleqatsiaq Peary assured.[13]

Another hunter in Qaanaaq, Qillaq Danielsen, told the researchers that killer whales had previously been killed in Qaanaaq. He knew that in the 1980s a group of hunters shot and killed a whale, but like Peary he added that now killer whales and hunters help each other.

By a local law, narwhals in Qaanaaq can only be hunted from kayaks, and the hunters are highly skilled at this specialized hunt. Before they kill the narwhals with a shot from a rifle, they need to secure it with a harpoon. The harpoons are handheld and carry a line back to the man in the kayak. Not only is this the traditional way, but it also ensures that the animal doesn't sink once dead. A narwhal in the open waters has more ways to escape a hunter in a kayak than a narwhal chased into the shallows by a killer whale.

I struggle to come to terms with Greenlandic people's killer whale hunts. To me, it seems poorly justified to kill animals that are not eaten but merely hunted for dog food or to protect another species that the hunters take; in Greenland, that is seals or narwhals, and in Norway, that was herring some decades ago. Not only are the ethics of this debatable, but the biological value of it has also been refuted as the scientific justification for culls is usually highly lacking.[14] In the case of seals and narwhals, there is evidence from other places, even within Greenland, of their healthy and strong populations even when killer whales predate on them. The issue of animals that are not retrieved but sink (or disappear wounded) is another concern. Researchers in Greenland estimate that as many as 75% of hunted killer whales are not retrieved. On land, most hunters would probably agree that this would be an unacceptable and unethical result of a hunt, but for some reason, different rules seem to apply in the water, where it is more difficult to see what is going on.

As I prepare to leave Tasiilaq, I can't say that I learned everything I came for. The Greenlandic hunt leaves many questions unanswered. It is a widely accepted belief that Inuit have a special relationship with nature and that this is automatically sustainable. Yet I can't help but wonder if the idea of the Inuit as a special kind of hunter in a pact with nature is one of those persistent stereotypes, grounded in traditions and some truths but not totally accurate today. Are the Greenlandic hunters different than any other hunters selecting which prey is most desirable to hunt and which species they prefer not to be in their backyard? Is this not exactly like the Danish hunters I have met who shoot the black birds—the crows, the rooks, and the ravens—because they compete for the pheasant chicks the hunters want to catch?

While serving as minister of culture, education, science, and church in the Greenlandic Government, Henriette Rasmussen, who is Inuit herself, pondered this as well:

> *Our people are not accustomed to the idea that efforts be made to keep hunting and fishing on a sustainable level, thereby implying that it is not already so. Tradition says, yes, of course, that kind of activity is sustainable! It has sustained us for thousands of years, it will do so in our time as well, and in that of our children and grandchildren! What else?*

She then goes on to answer the question she raises herself: "*Sustainable use of nature has become a catchword. But, not so 'sustainable development!' That, one must admit, is a much more difficult concept to deal with.*"[15]

Culture and tradition weigh heavily in the lives of modern Inuits as they struggle to honor their heritage and carve a living out of the harsh conditions in the Arctic. Survival is perhaps not the same issue to modern Inuit as it was to their ancestors, but it is not only a question of getting enough to eat. In the supermarket in Tasiilaq I saw most of the same things that I can buy in a supermarket in Denmark, but it was very expensive. Customers must pay the cost of transporting all the household goods that most of us take for granted on cargo ships from Denmark to Greenland. No vegetables or fruits grow in Greenland (except for tiny black crowberries in the tundra), and there is and cannot be any agriculture since ice covers most of the land. It is no surprise that hunting is at the core of Inuit culture and traditions and is an important source of identity and self-esteem. A successful hunter enjoys the admiration of many and usually has a high status in his community.

In parallel fashion, different cultures and traditions are exactly what make killer whales so fascinating, as different populations maintain different hunting habits, food preferences, vocal repertories, and other socially inherited characteristics. But it is also what makes them vulnerable because unique cultural traits can be lost if a small population goes extinct.

Nowhere in the world are killer whales common. Their populations are usually small, sometimes even so small that the risk of extinction is imminent. Because killer whales occur in distinct ecotypes in many places, what to the untrained eye looks like one strong population may be several separate, smaller, and more vulnerable groups. These ecotypes are well studied in some places around the world, like the Pacific Northwest and in Kamchatka where fish-eating killer whales occur in the same waters as marine mammal–eating killer whales. In Norway, too, both fish-eating and mammal-eating ecotypes of killer whales have been identified, although it seems that the marine mammal ecotype are unusually flexible in their dietary preferences and will occasionally eat fish.[16] And there may be other populations not yet identified in Norway.

Long after researchers in Canada had established what kind of killer whales lived in the waters in western Canada, they discovered a population of killer whales in offshore waters that eats fish but has specialized in sharks and other bony fish rather than salmon. In Antarctica, it is believed that there may be as many as five different killer whale ecotypes, all separated because of their varying dietary preferences and group-specific behaviors.

Since there are many places where killer whales have not been deeply studied, there may be many more ecotypes out there. Each of these cultures is at risk of disappearing if a subpopulation goes extinct. Too little is known about the killer whales off Tasiilaq. They may very well belong to a small and vulnerable population of killer whales with a separate but not yet described set of traditions.

Stepping into this discussion became exactly as difficult as I feared it would be. Tobias Ignatiussen and Gedion Ignatiussen met me with friendliness and hospitality, and they both generously shared their experiences and thoughts about the situation with me. It makes me feel ungrateful and rude to reciprocate their hospitality and generosity with any sort of criticism. I appreciate that their culture is of value, not only to themselves but to all who believe in the value of cultural diversity, and I hope that they find a way of coexisting with killer whales that is in line with their culture and way of living.

On my last day in Tasiilaq, I didn't need to look out of the window to tell that the wind had picked up. The violent weather made the wooden building where I stayed shudder and creak as if it was a ship in a storm. Outside, the ravens played in the wind like they had nothing to worry about. And maybe they don't. Although new executive orders from the government have allowed for unregulated hunting of ravens, foxes, and wolves year-round, the ravens, unlike the killer whales, are rarely hunted.

CHAPTER 11

Family Matters

Five thousand kilometers away from Tasiilaq, practically on the opposite side of the globe, the attitude among the indigenous groups in the Pacific Northwest is entirely different from the attitude among the Inuit hunters. When the orphaned killer whale Springer returned in the summer of 2002 to the area where she was born, she was welcomed by song from the local 'Namgis tribe who had gone out in their traditional canoes to greet her. To them, she and her kin are beloved family members.

Springer was first sighted June 9, 2000. At that first observation she was a little calf following her mother as closely as if she was on a tether. The mother was in the A4 pod, which belongs to the Northern Residents, a population of killer whales that is found from central Vancouver Island to southeastern Alaska. The Northern Resident populations are named alphabetically with letters A through I, and their southern cousins are J through L. The A4 pod was named after a large, now long dead, bull called

A4. Here in the Pacific Northwest, in the three separate populations that inhabit these waters, killer whales have been studied so intensely and for so long that researchers know each individual and its life history.

Later that same year, in September 2000, Springer and her mother were resighted, along with the rest of their family, by researchers in southern Alaska. But the year after, when the family returned to the waters off Vancouver Island, neither Springer nor her mother were seen, despite the researchers having had many encounters with the A4 family during 2001. Family ties are so strong in killer whales that this was taken as a sign that the two had died.

Indeed, the mother has never been seen again, but Springer appeared most unexpectedly in the harbor of Seattle in January 2002, more than 400 km south of the usual habitat of her family. Crew on the ferry to Vashon Island southwest of Seattle noticed the little killer whale that seemed unimpressed by the size of the ferry and was eager for contact.

It is not unusual to see killer whales close to the harbor of Seattle, but this sighting was peculiar because the whale was on its own. When a killer whale is encountered alone, it is usually just for a short period of time before the rest of the family is also discovered and you see all of them swimming together again. But there was no family around this time. Day after day the whale continued to be alone, and it approached boats and people, seemingly looking for company and interaction.

The appearance of the little whale attracted attention from enthusiastic whale-lovers but was also a cause of concern. Just a bit more than 3 m long, she was clearly not much more than a baby, and she did not look healthy. Underweight with discolored and flaky skin, her habit of approaching boats and interacting with people was deeply worrying. However cute it was to some of the onlookers, this kind of behavior jeopardized her chances of survival. A whale that has replaced their normal social behavior with bonds to humans has a much higher risk of being fatally struck by a boat or becoming dependent on people for contact and food.

In the beginning, nobody knew that the little whale was Springer, and researchers were unsure where she had come from. To recognize individuals, researchers look for unique patterns and marks found in a whale's grey dorsal saddle and the back fin, and occasionally in the form of the white eyepatch. In young calves, however, these marks are seldom as pronounced as in older animals. In the case of the young whale in the Seattle harbor, the most obvious guess would be that she belonged to the Southern Resident population, the small population of fish-eating killer whales that is mainly distributed in the Salish Sea between Seattle and Vancouver Island. But like the Northern Residents, these whales had been studied for decades by this time, so all the whales were well known, and the little whale didn't fit with any known individuals in the Southern Resident population.

A sound recording of the whale's vocalizations was the clue to her identification. Many of the calls were indistinctive vocalizations, or "baby-talk," but a few calls were clear and distinct and placed the whale securely in the A4 family in the Northern Resident community. Some calls are shared between many different pods, but all pods also have their own "dialect," calls that are unique to them and can therefore be used for pod identification. Although Springer's mother was considered dead, the rest of her family, including her grandmother and great aunt, were still alive in the A4 family, and the Seattle-based conservation group, Orca Conservancy, immediately suggested that an effort should be made to reunite Springer with her family.

It was a hard decision to make. An attempt to bring Springer back to her family would not only demand close cooperation between the authorities on both sides of the border but also agreement between different conservation organizations, researchers, and dedicated whale enthusiasts, many of whom had access to the media. Some feared that Springer could easily end up captive for life in a tank in a marine park. Others feared her imminent demise if a reunion was not tried. Nobody knew how an attempt to reunite Springer with her family would end. Even if the capture and transport of her went smoothly, there was no guarantee that she would connect with her family again. Plus, the operation would cost a ton of money. Previously, marine parks and research facilities with captive dolphins had tried releasing their captive animals to the wild, sometimes with success, other times ending with an animal's death. Most of these releases had involved bottlenose dolphins; the catching and relocation release of a killer whale had never been tried before.

Luckily for Springer, people loved her. The longer she stayed in the Seattle harbor, the more she became the darling of the public. It became clear that if nothing was done, she would become dependent on people for both food and social relations, meaning it would no longer be possible for her to return to a life in the wild. At the end of May 2002, after almost five months in the Seattle harbor, the National Oceanic and Atmospheric Administration (NOAA) decided that an attempt should be made to reunite her with her family in their home waters. NOAA coordinated the effort, working with several conservation groups: the American Cetacean Society, Orca Conservancy, OrcaLab, and the Whale Museum in Friday Harbor, as well as the Vancouver Aquarium and Department of Fisheries and Oceans in Canada.

Because Springer was fond of playing with sticks and interacting with people, she was easy to approach and keep stationary long enough to be captured. All it took was to lure her with a stick in between two dinghies and then stroke her while she was led into a specially prepared sling. A 65-foot crane barge lifted her into in a sea pen near Manchester on the Kitsap Peninsula in Washington State. Here she was closely monitored and treated for a skin disease as well as internal parasites. And she was fed, too, live salmon that she had to catch herself so she would not become more dependent on humans than necessary. After a few weeks, she grew healthier and stronger and, in mid-July, her caregivers gave the signal. She was ready for release. On July 12, 2002, the big crane lifted Springer again, this time onto a fast 144-foot catamaran that sped north, back to her native waters.

Another sea pen had been prepared for her in Canadian waters, close to the coast of Hanson Island at the northern entrance to Johnstone Strait. Johnstone Strait is a favorite place for many of the killer whale pods in the Northern Resident community, including the A4s, Springer's natal pod. Nobody knew how long it would be until her immediate family swam close enough that Springer could be released from her sea pen to join them. Fortunately, they didn't need to worry for long. The A4s came by the very first day that Springer was in the pen. And she must have heard them and recognized their underwater calls because she immediately started spy-hopping and repeatedly pushing the netting of the sea pen with her head. A journalist reported that Lance Barrett-Lennard, a scientist from the Vancouver Aquarium who was monitoring her activities, said, "*Her calls were so loud they practically blew our earphones off.*"[1]

Still, the people involved in the release, the researchers, the veterinarians, the staff from Vancouver Aquarium, and lots of other experts who had been advisors in the relocation process hesitated, asking themselves if it was premature to open the gates on the very first day. In the end, they decided not to and it took hours before Springer calmed down again. Even after her family had left the area, she continued to jump and spy-hop and cry out. The A4s came back the very next day, and like the day before, Springer grew agitated—or excited—when her family passed by her enclosure. This time nobody thought twice.

They immediately lowered the net so Springer could swim out. All around the sea pen and in the waters outside Hanson Island, a flotilla of boats waited to watch the release. Among them were traditional canoes from nearby Alert Bay, the nucleus of the 'Namgis Nation, in contrasting black and red. Many of the people were dressed in traditional regalia, including button blanket capes in the same contrasting colors as the canoes, while some wore cedar bark headdresses and other ceremonial objects. One of the canoes unloaded two men, Chief Bill Cranmer and Ernest Alfred, onto the gangway of the sea pen before the release. Here they sang a traditional song for Springer, wishing her a happy reunion with her family.

When the net went down, Springer bolted. She swam away from the enclosure at high speed to catch up with her family, but then a branch in the water caught her interest and she started to play with that instead, like a toddler whose attention is immediately distracted if there is a toy lying in their path. The onlookers held their breath. Would she find her family, and would they accept her?

They didn't get the answer straightaway. At the release, Springer was equipped with suction-cup radio tags, allowing the researchers in the area to follow her whereabouts. The day after the release she was observed with part of the A4 family in the Robson Bight Ecological Reserve, a small bay in Johnstone Strait famous for its pebble beaches, on which the whales rub themselves by swimming over the pebbles in shallow water. In the following weeks, Springer was seen with different groupings of her family many times, but she appeared to lag behind them when they were swimming fast, and researchers worried that she did not have the strength to keep up. Other times she was alone and resumed her worrying habit of approaching boats—a behavior that could potentially become fatal. But there were many times Springer also seemed to reunite with her family, and after the first wobbly weeks, she swam with them regularly. It looked promising.

By fall, the killer whales left the area around Johnstone Strait, as they do every year, to spend the winter in more offshore areas. It would be at least six months before there was a chance to see whether the ties Springer had reknotted would hold and whether she could survive a winter in the wild. According to a local newspaper, John Ford from the Department of Fisheries and Oceans in Canada, one of the researchers involved in her release, said "*This is the real test whether she makes it through the winter. We will just have to wait and see.*"[2]

Approximately eight months later, on July 10, 2003, Springer was identified with a large group of killer whales north of Vancouver Island. Her story had ended well; she was still with her family. She was reported to be in excellent condition, and her reunion with her own family is still the only successful reintegration into the wild of an orphan killer whale. Since then she has been resighted annually in Johnstone Strait, and in 2013 and 2017 she even returned with calves of her own.

In 2022, 20 years after Springer's release, I visited Ernest Alfred to hear his story and to learn more about his involvement with killer whales and conservation. During a visit with friends in Seattle, I drive to Port McNeil at the north end of Vancouver Island to take the 40-minute car ferry to Alert Bay on Cormorant Island to meet him. The drive is meditative. Except for a few towns on the way—Nanaimo, Parkville, and Campbell River, which appear as exit signs on the way—the land I pass is deserted of people and almost drained of colors. A charcoal black road runs like a solidified stream through a dark green landscape of cedars, hemlocks, spruce, and pine trees. A cloak of clouds hangs low, hiding the treetops in a grey fog.

The first white explorers to reach the Pacific Northwest raved about the region's riches. Trees as tall as towers; sea otters, sea lions, seals, and whales everywhere; beavers and bears bouncing from land to sea; and salmon in such copious amounts that you could walk across a river on their undulating backs. The exploitation that followed was as inevitable as rocks rolling down a mountainside. The story of the near extinction of the sea otters with their luxurious fur is the most well-known, but other species were also targeted and populations of seals, sea lions, fur seals, and whales were seriously depleted. Loggers cut down the lush forests and floated the timber down rivers and into the sea to be processed at lumber mills.

Most of Vancouver Island is covered with temperate rainforest but more than 90% of the precious old growth forest has vanished; giants that were many hundreds of years old have been cut down and sold as timber. The trees I pass on the drive are clearly not old growth; they are skinny, of middling height, and their branches lack the mosses and lichens that cover those in the ancient forest. But the trees are still plenty, and logging continues to be an important industry. On mountainsides I see the telltale signs of a clear cut; the patches of bare land, irregular and raw, like scars from a bad burn.

When I started the drive, the trees lining both sides of the road formed continuous green walls. Later, when it starts to rain and my field of vision narrows to the part of the landscape I glimpse through the rhythmic motion of the windshield wipers, the trees have all but disappeared in the grayout. It rains for hours. The place in North America with the highest average rainfall per year is Vancouver Island. There is a reason that rainforests grow here.

In Port McNeil, the houses facing the little harbor display "Forestry Supports My Family" signs in their windows and a few fishing boats sit in the harbor, while men in pickup trucks stop in the road to chat. The sky miraculously clears during the ferry ride to Alert Bay, so when I arrive there are patches of blue sky and a few streaks of sunlight. Alert Bay is a small community in a traditional 'Namgis territory; only about 1,200 people live there, half of them First Nations. Until the 1980s, they were called Kwakiutl, but today they prefer Kwakwaka'wakw Nation, which means those who speak Kwak'wala. The Kwakwaka'wakw consists of many different tribes, each with their own geographical roots. Alert Bay is now the heart of the 'Namgis tribe. A welcome arch, carved in cedar, salutes visitors with a large "Home of the Killer Whale" figure designed in the typical Northwest Coast art style, with curved lines and ovoid shapes painted in red, green, black, and white.

In the harbor, the pale yellow Seine Boat Inn, the Orca Inn (featuring both a pub and a restaurant), and the Nimpkish Hotel (with a sundeck facing the sea) compete for space with a handful of dock houses constructed of corrugated iron on stilts. The roof is gone from one of the dock houses and the windows are unhinged and glassless. The door is blocked with a piece of plywood. Driftwood piles up along the shore, but not the kind you would find on beaches in Denmark or Norway and bring home to decorate a windowsill. Most of this driftwood consists of whole gigantic tree trunks with their roots still attached. Ramming into this kind of driftwood with a sailing boat on a dark night could easily be fatal.

The most common bird sound in Alert Bay is not the chatter of sparrows or the squawks of seagulls but the high-pitched screech of the bald eagle, which doesn't seem quite majestic enough for the bird's royal rank. Like pigeons in a city, they are everywhere. In a tree next to the Big House (the 'Namgis Nation's ceremonial community house), eight or nine eagles perch, their white heads conspicuously contrasting the dark green of the cedar. Six other eagles, younger and more scruffy-looking, sit on a nearby roof. Bald eagles don't get their white head and tail feathers until they're around five years old. I pass one that is entangled in some brambles at the ground, but it manages to free itself when I approach. Six or seven soar in the sky.

When I turn a corner, it's the same scene. I reckon that there must be at least a hundred bald eagles in the bay, maybe even several hundred. The eagles are accompanied (and bullied) by ravens, which are almost as plentiful. Like the ravens in Tasiilaq in Greenland, they are bold and self-assured, approaching people and eagles with equal confidence in their ability to escape with a brash aerial maneuver.

At the southern end of Alert Bay, I find the 'Namgis burial ground. Like other graveyards, there are graves with wooden crosses and names inscribed, but more than 30 totem poles rising from the ground makes it unlike any other graveyard I have seen. From a distance, the totem poles look like dead trees, tall and lifeless snags left after a windfall or a forest fire—until I see the figures carved in them. Silently they all face the sea, marking the death of chiefs and community leaders of the past. Some of the totem poles are old and withered; others look new and are painted with bright colors. A few are so old that they have fallen over and now lie like a tree trunk in the woods, rotting and overgrown with moss. Both the old and the new poles are full of symbolic figures, otherworldly characters and spirits from the woods and the sea with fierce eyes and teeth and fangs, as if they are ready to grab any passing soul. Some are readily identified, like the eagle with outspread wings and a large, curved beak, and the numerous killer whales with their crowded mouthfuls of teeth and their backfins sticking out perpendicular to the pole. Other figures look like supernatural characters, combining human and animal features.

I meet up with Ernest, hereditary chief of the Tlowit'sis people and elected councillor of the 'Namgis First Nation, in the small guesthouse where I am staying. He is a young man with short dark hair, dark smiley eyes, and an energetic spring in his step. Unlike many of the pictures I have seen of him before my arrival, where he was dressed in the traditional regalia of his tribe (a red and black blanket decorated with buttons, a cedar bark neck ring, and a cedar headpiece with ermine pelts), he is wearing a plain black t-shirt with a recycle logo, faded jeans, and a thin down jacket. He explains that the traditional regalia are used both in ceremonies and official and public events, such as the release of Springer. And the song that he and Chief Bill Cranmer sang for Springer was in Kwak'wala, a language that is spoken by fewer than 3% of Kwakwaka'wakw today.

"It was more of a chant actually," Ernest clarifies when I ask about the meaning. "It was like a prayer wishing her well, hoping she would find her family."

It is not only Springer that Ernest sings for. Each summer when he encounters the killer whales in the area for the first time, he sings a welcome song to them. The whales are his relatives, his ancestors. Traditions run deep in his family.

"We have a really direct connection with the killer whales," he explains. The killer whale is an important crest animal in his family, who can trace the origin of the whales' importance to their family to his great-great-grandfather who once lived further north, near Fort Rupert on Vancouver Island. A crest animal signifies kinship, group membership, and identity.

"This great-great-grandfather had his home up there with the Kwakiutl," Ernest says. "At some point a spell was cast on him and he fell very ill." Stricken by the spell, his grandfather went down to the ocean to pray. When a group of killer whales came by, he took water in his mouth and blew it toward one of the whales and asked it for help. Later the whale beached itself in front of him, sacrificing its life, and then his grandfather recovered.

"That's how the killer whale became our crest animal," he says. "I really am a whale person. The whales are my relatives. That is how we see them in my culture."

When I ask about the meaning of the welcome song he sings to the killer whales, he pushes back his chair and starts singing it instead of explaining the words. His voice fills the room, and it is as if a transformation takes place in him and our surroundings. He looks up as he sings and chants the words in his native Kwak'wala tongue; there is not the slightest moment of awkwardness or embarrassment in the situation. The regalia, the drum, and the cedar headpiece with the dangling ermine pelts are not necessary to create this moment of solemnity and gravitas. I understand that this is who he is. He may be a schoolteacher, an activist, and a Canadian citizen, but first and foremost he is a leader of his community, and this is his language.

The words of the songs are *la'am's hutlila gaxan*, which means "You will listen to me." Ernest explains that it is the same song and the same words that are used at every ceremony in the tribe when family members are welcomed one by one as they enter the Big House. The killer whales are family, too.

"I believe they recognize me," he says. "I believe they recognize the song. I believe they know who I am, because I have been singing this song year after year." He explains that sometimes the killer whales change behavior when he sings. They become excited and start spy-hopping, rolling on their sides, or breaching.

In 2009, Ernest took on duty as a warden in the Robson Bight Ecological Reserve, the bay where killer whales from the northern community come to rub themselves on the pebbles at the bottom. The rubbing is a highly specialized behavior that is unique to the Northern Residents and is another behavior like the pod-specific dialects and food specializations that are considered cultural traits. In the following years, Ernest became increasingly aware and worried about the decline of the salmon in Canadian waters, a problem for killer whales and many other species, as well as for people depending on salmon for food and income.

Many biologists say that salmon is the keystone species of the Pacific Northwest. However, what exactly defines a keystone species is not always clear. Some define it as a species that has a disproportionately large effect on its natural environment relative to its abundance; others simply classify it as a species that is a vital component of an ecosystem, holding it together because of its importance. By the last definition, salmon is certainly a keystone species. It is a crucial species in the food web in the Pacific Northwest where it is—or was—present in such vast numbers that many other species depend on it to thrive and proliferate, or even exist. More than 100 species depend on salmon for their own existence, at least for part of their lives. Among these are charismatic species, such as grizzly bears, black bears, bald eagles, and killer whales.[3]

Salmon is, of course, important to people as well. An indigenous woman I meet in a small shop in Alert Bay tells me that the only time she had ever been away from her home in Alert Bay, she saw a box of salmon for sale in a shop in Vancouver. She was shocked both to learn that salmon came in boxes and that the price was $25.

"When I came home for Christmas, I gave my father a big hug," she laughs. "I realized that in all those years when I was a child, he had been providing for me and the rest of the family with his boat, his hands, and the labor he put into getting fresh food on the table." She goes on telling me that for many in the community, the sea is where food comes from, and salmon is the bread and butter of their livelihood. "Or at least it used to be," she finishes.

Living from the sea has become increasingly difficult as many salmon stocks are considered critically endangered. In the Pacific Northwest, a majority of the different local stocks of the five most important salmon species—Chinook, sockeye, pink, coho and chum—are either considered threatened or endangered. In a straight-talking 2020 Washington State report, the message is not shrouded in ambiguity: "*Too many salmon remain on the brink of extinction. And time is running out*," it reads. A heap of factors contributing to the decline are listed in the report, including the warming of rivers and oceans, habitat destruction, landfills, hydropower, and dams.[4]

For seven years, Ernest watched over the killer whales in the Robson Bight Ecological Reserve and made sure that they were not approached by boaters or others who could disturb them. But increasingly he brooded over the problem of the declining salmon, and in 2018 his thoughts about conservation work prompted him to take swift action. On August 24, 2018, he occupied Swanson Island, a small, 4 km long island in the Broughton Archipelago east of Vancouver Island in Canada. Here, the Norwegian company Marine Harvest had a system of open-net fish farms.

"I had become increasingly worried about our natural salmon populations," Ernest says, "and frustrated with the fish farms." He believed that the fish farms were a principal reason for the decline of the natural salmon populations. In the first days of the occupation, he took shelter in a tent and announced that he was not leaving until the fish farms were removed. Later he moved into a cottage on the island that had been erected to accommodate the men working on the adjacent salmon farm. He received support for his mission from activists, conservationists, and First Nation communities in the area.

At first, the few workers on the salmon farm ignored Ernest—but after a few days, their geniality ran out and they asked him to leave.

"You are trespassing," they told him.

Ernest replied that it wasn't him but *they* who were trespassing. It would be another 284 days before he returned to his home in Alert Bay. During that time, he posted updates every day on social media about his observations from the fish farm. He chuckles when he admits, "One reason for choosing Swanson Island for the protest was that it was part of the 'Namgis's traditional territory, but I also needed good cell phone coverage. And Swanson Island is one of a few islands around here that has that."

He used social media platforms to raise awareness about what was happening on the fish farm while educating his followers about the parasites that infested the farmed salmon and the virus that farmed salmon spread to wild salmon. Ernest also shared his deep concern about the pollutants, like prescription medicine, washing from their sewage system into the sea, not to mention the genetic pollution caused by escaped salmon, which was not a local species but imported Atlantic salmon.

The science largely agrees with Ernest on the issue of salmon farming. In a study from 2008, researchers found that wild salmon populations that have been exposed to salmon farms on average suffer a reduction in survival or abundance of more than 50%, compared to populations that have not been close to salmon farms.[5]

When Ernest left Swanson Island almost a year later, nothing much had happened in terms of immediate action. But in December 2019, Prime Minister Justin Trudeau instructed his fisheries minister Bernadette Jordan to work out a plan to end open net fish farming in Canada. The first steps have been taken. In December 2020, Bernadette Jordan announced the closure of 19 salmon farms in the Discovery Islands, about halfway up Vancouver Island's eastern coast. But that was just a fraction of the hundreds of fish farms in British Columbia, and the owners, which in most cases are Norwegian companies with licenses to the farms, were not planning to wrap up their business without resistance. They announced that they are taking the government's orders to court.

On his return to Alert Bay, Ernest received gratitude and applause from his community. He had raised the issue of the negative impact of the salmon farms on wild salmon and the marine environment and pushed hard for a solution, reaching thousands of people and making sure that the message was heard all the way up to the government. But the fight didn't come without personal costs. When he started the occupation, he left his regular job in Alert Bay, earning no money in the meantime, and he also met hostility among people dependent on the fish farms. Like the people who put up the "Forestry Supports My Family" signs in their windows in Port McNeil, the salmon farming employees also fought the conservationists. Ernest received hate mail and threats and is still emotionally scarred by the turbulence that he became the center of.

In the summer of 2011, I went to the San Juan Islands. I was visiting my friends Mike Dougherty and Catherine DeNardo in Seattle. Together we had spent several happy years studying whales in Norwegian waters, and it was an easy decision to rent a sailing vessel to go on a cruise in the beautiful waters around the San Juan Islands. When we let go of the mooring lines in the harbor of Bellingham, north of Seattle, and set the course eastward toward the Strait of Georgia, we didn't expect to see killer whales. The area we were going to was at the northern range of the Southern Resident population of killer whales, and before departure we had checked with local whale-watching companies to see if there had been any sightings lately. We had been told no, it had been a few weeks.

We went anyway. We were looking for a week of peaceful sailing, and we would be perfectly happy with each other's company, the occasional bald eagle, and the seals and sea lions that we most certainly would find. The San Juan Islands is an archipelago positioned almost exactly at the border between the United States and Canada. Some of them, like Deadman Island, Goose Island, and Sucia and Matia Islands, are uninhabited and covered in cedars and Douglas firs. Others, like Lopez Island, Orcas Island, and San Juan Island, have a few small towns and are lined with holiday homes hidden between the characteristic, red-barked madrona trees overlooking the sea. Our destination was Patos, the northernmost island in the archipelago, which is only about 1 km long and one of the uninhabited islands.

As we neared Patos, we lowered a fishing line and tried to catch fish for dinner, but in the end we had to make do with the olives and pasta and wine we brought. We arrived at sundown and anchored in a little cove only a couple of meters from land. In the quiet summer night, I slept on the deck, and it was still very early in the morning when I woke up to the sound of a curious chatter and hissing. On the shoreline, a family of river otters paraded, squealing and squeaking nosily as if they were gossiping about something outrageous they had just witnessed. During breakfast, a harbor seal stuck its head out of the water, battling with a very large salmon. The fish lost the battle and the seal's triumph made us laugh at our own unsuccessful attempt at catching yesterday's dinner.

Before we had finished our morning coffee, Mike spotted two Dall's porpoises crossing the waters at the entrance to the cove. They darted across, fast and powerful like dolphins. We were following them with binoculars when we discovered suddenly the tall distinctive fin of a male killer whale cutting its way through the flat water about a mile further out. We quickly cleared the breakfast table, pulled up the anchor, and were off in no time. Once we were out in the open water, we found out that there were killer whales all over the place. They were dispersed widely in Boundary Pass, the strait between the San Juan Islands on the US side of the border and the Southern Gulf Islands on the Canadian side, some moving in small groups, others alone, but all of them heading south. They were in no hurry.

We ended up spending the morning following them slowly while keeping a distance at the same time. They behaved slightly differently than their Norwegian cousins. Feeding on schooling herring is much more of a coordinated effort, usually involving a whole pod of killer whales, and therefore it is more typical to see Norwegian killer whales in tighter groups. The American killer whales were also making many more spy-hops, breaches, and bellyflops at the surface than we see in Norway. This seemingly exuberant aerial behavior is a distinctive of the Southern Residents, known for being very active at the surface, and considered a cultural characteristic—something they do because it has become ingrained in their behavioral repertoire through generations and generations of social learning. It was impossible for us to count the whales, but we estimated that there were at least 60, possibly more. That means that probably all three groups of the Southern Residents, the so-called J, K, and L pods, were present in what is called a "super pod" gathering.

At the time of our visit, there were 88 whales in the Southern Resident population, but in February 2022, a bit more than a decade later, the number has gone down to 73. Since 1996, censuses of this population have mainly gone downhill, and in 2006, the Southern Resident killer whales were listed as endangered under the US Endangered Species Act. A surge of births in 2015 gave cause for hope but not all survived, and in the following years, no new calves were born. Meanwhile, the older whales continued to die, so despite the newborns, the populations continued to shrink. The Northern Resident killer whales are doing better. Like the Southern Resident killer whales, they became listed under the Species at Risk Act in 2003 but "only" as threatened. Their population in 2022 is around 300 individuals.

Like killer whales elsewhere in the world, the Northern and Southern Resident populations are specialists, preferring Chinook salmon, which makes up around 70% of their diet.[6] Chinook is the biggest and fattest of the different salmon species, and for that reason is often called king salmon. It is also the least common. Therefore, when in the mid-1990s the Chinook salmon in the Pacific Northwest suddenly was even less common than usual and stayed so for more than five years, it had a detrimental effect on both killer whale populations. Each year, the killer whale researchers went out, took ID pictures, and matched them with their ID catalogues, and they found that there were individuals missing. Mortality increased rapidly, and both populations suffered substantial losses. The Southern Resident population in those years went down 17% and the Northern Residents dropped by 8%.[7] Another consequence of the salmon decline can still be seen in the killer whales that were young and growing at the time of the shortage of food. Their growth was stunted, and they are on average 40 cm shorter than older killer whales in the same population.[8]

When the Chinook salmon bounced back around 2003, the killer whale populations stabilized again—at least in the north. The Northern Resident population continued to grow,[9] and Springer's two calves, born in 2013 and 2017 in the A4 pod, now contribute to the survival of the Northern Resident killer whale population. But the future for the Southern Resident killer whales looks dire. Since 2011, this population has experienced an almost steady decline. In fact, the exact same can be said about them as was said about the salmon in the Pacific Northwest: "*They are on the brink of extinction. And time is running out.*"[10]

Ironically, where killer whales are probably most threatened today is not where they are hunted, like in Greenland, or captured, like they were in Russia (the ban on live capture has been extended till 2023)—it is in the Pacific Northwest where they are loved and treasured. Springer's rescue was not motivated by a conservation concern. She was not captured, treated, and reunited with her family because her future contribution to the growth of the Northern Resident killer whale community could become essential for the population's long-term survival. She was saved because she was loved. The affection of thousands of people brought her back to her family. Those who didn't participate directly in the rescue mission donated money for the operation, or they helped fundraise the complicated effort of returning her to her family.

But all the love in the world is not enough to help the Southern Residents, and saving one orphan whale is a trivial task compared to changing the conditions that threaten these whales. Both NOAA in the United States and Fisheries and Oceans Canada have developed extensive recovery plans for the killer whales in their regions. They mostly agree on what is needed: the whales must have enough to eat, the level of chemical and biological pollutants in the ocean must be curtailed, and noise pollution from loud human activities—from cargo ships, dredging, drilling, construction, seismic testing, and military sonar—must be controlled. Right in the Southern Resident whales' range are some of the busiest, and loudest, shipping lanes in North America.

The big difference between saving Springer and saving a whole population of killer whales becomes obvious when one reads these plans. Among the steps described in the recovery plans are reduction of vessel traffic, removal of hydrodynamic dams, reconstruction of freshwater ecosystems, limits on fishing and aquaculture, and restrictions on movement of leisure boats. All are political hot potatoes. Perhaps Fisheries and Oceans Canada is not too sure about the political success of the plan either. At least they conclude, "*Due to its small size, the Southern Resident Killer Whale population will be particularly vulnerable to catastrophic events and continues to have a high risk of extinction during this period.*"[11]

Or perhaps they are underestimating the love for killer whales in the Pacific Northwest. Maybe people are willing to welcome all these political changes to protect what is, in the hearts of so many, the most iconic species in their part of the world.

CHAPTER 12

Cut in Stone

In the years before killer whales moved farther north, I brought my father, my twin sister, her two sons, and my own daughter to Tysfjord to share with them some of the things I had spent so many years doing. On an utterly cold day, we crisscrossed the fjord, watching killer whales chasing herring and white-tailed eagles swooping down to snatch the leftovers at the surface. It was late November and when we returned to the harbor most of the light had already faded. Instead of going back to the warm cabin where we stayed, I convinced my family that we should go on one more adventure.

"It is not so far," I beseeched them, "and it is something you can only see here."

I wanted to take them to the legendary rock-carving site at Leiknes in Tysfjord. We got into the car and followed the narrow road that winded its way along the rocky shore of the fjord. About 10 km farther on we parked on the curb next to a humble road sign telling occasional visitors how to find their way to the rock carvings. From there we followed a footpath that climbed gently uphill to the site. The path was relatively easy to walk, which was a good thing because my father didn't walk very well. He used a cane and had very little balance, and I hoped that the path would not be too difficult for him. He looked brittle, like a dry stick. My mother had died a year earlier, and this trip was also an opportunity for all of us to do something together after a difficult time.

The light was low, which was also a good thing because the carvings are best seen at low light. They were discovered in 1912 by a local boy.[1] He was then about 10 years old and lived on a nearby farm. The area behind the farm was open woodland on smooth sloping rocks, a good place for children to explore and play. He and his friend discovered that the rocks were full of animal figures, and they called it Dyreberget (Animal Mountain). For a time, it was their favorite playground.

When a new road was being planned a few years later, one of the engineers heard the local people talk about Dyreberget with the animal figures and alerted museum authorities to the phenomenon. It was then officially described and is today valued as one of the finest examples of prehistoric art in Norway. Few people go there, even though local tourism has blossomed the last decade. Apparently, the people who come favor the wildlife in nature over the kind that is embedded in rocks.

The path meandered its way through the low bushes, snaking between the few birch and pine trees in the boggy area. It was wet from the snow that fell a few days ago and then melted again, leaving everything muddy. I was relieved it was not frosty and covered in ice, which would have made it impossible for my father. As we started to climb, the hike got a bit tricky, especially where we had to walk on sloping rock faces. They were both slippery and difficult to balance, even for the more nimble of us. I kept an eye on my father as he staggered over the rocks, but he smiled back to me that he was okay. Even out in the open there was no wind. In the quiet terrain, we heard two ravens calling to each other just before flying over our heads, their raucous voices bouncing back and forth between them. After about half an hour's walk, the vegetation thinned out and opened onto an expanse of slick rock surfaces.

We had reached the site of the rock carvings and saw the first, a beautiful couple of swans. Or maybe they were geese; it was a bit hard to tell. Their rounded bodies continued into long, graceful necks and small heads, and the neck was precisely a bit too long for a goose and a bit too short for a swan. They were made with a simple line as if the artist were told he could not lift his paintbrush and had to make them in one stroke. Their bodies were slightly overlapping with the two long necks next to each other, both birds looking the same way. The artist had used a natural quartz ore in the rock to simulate the waterline the swans were swimming in.

Unlike most rock art elsewhere in Norway and in other places in the world, the figures at Dyreberget are not carvings but polished lines. It's astonishing that the lines are still visible, even now, thousands of years since they were made. Apparently the glossy surface makes it difficult for lichens to grow there, so the lines stay lighter than the surrounding rocks, which are weathered differently. The polishing is also why the lines are best visible in low light. With the sun at a low angle, the lines are seen as narrow, slightly glossy contours. This special technique of decorating rocks is only used in seven other places, all of them found close to this site in northern Norway. Some archaeologists think that this type of rock art is the earliest form and that it preceded the more common rock carvings, where the artist actually carved into the rock with an instrument. They are estimated to be about 9,000 years old and are thereby among the oldest of open-air rock art in Europe.

It is almost impossible to imagine people living here so long ago in this cold and harsh environment without any of the modern commodities that we depend so much on today. They were Stone Age hunter-gatherers making a living from the sea and the land. Despite the high latitude, the coast was and still is ice-free year-round, and they probably lived there, at the boundary between land and sea, to make the most out of both elements.

Many rock carvings and drawings are found in places where it would be natural to cross a fjord or a bay and where there would have been plentiful supplies of fish and other types of food from the sea. Such sites also had access to inland game and wildlife. The rock art in many of these places are quite far from the shore today, as our arduous walk had just proved, but at the time when they were made, they would have been at sea level. Almost 10,000 years ago, the last ice age had just ceased, and the ice caps all over the Northern Hemisphere were retreating. Without the weight of the ice, newly uncovered land rose considerably and places that were once at the shore are now 40–50 m above sea level. Therefore, most of the rock art found in this area is now inland.

We continued a bit uphill from the swans to find the area with most of the animals some hundred meters farther up. When we arrived, we discovered that it was quite a large area, bigger than a tennis court, and we had to walk around it to find all the figures in it. As we walked on the rocks, we bent down to try to get a glimpse of the lines from different angles, just like trying to see a photograph in an old-fashioned film negative, where you have to tip the film to let it catch light at different angles. In this way, we discovered moose, reindeer, bears, a hare, and a big killer whale.

Some of the figures seemed to be left half done; just the head of a moose or the hind parts of a deer, and some of them were overlapping each other. They were all life-sized depictions and made in the same graceful style as the swans. The faint contours outlining their bodies seemed to be made in one stroke, making the animals appear elegant and willowy. They were all facing in the same direction, too, as if they were a herd on the move. One of the reindeers had its head turned, looking over its back in a way that was graceful and natural at the same time.

Like the other animals, the killer whale was drawn to actual size, more than 7 m long. It was obviously a male killer whale with a tall back fin. Oddly, though, the fin was curved the wrong way, bending forward instead of backward. Regardless, it is impossible not to wonder at the fact that even then, almost 10,000 years ago, killer whales were roaming these seas.

The killer whale at the Leiknes site is the only killer whale portrayed in prehistoric rock art in Europe, but there are a few other places in the world where killer whales are pictured in rock carvings. One of them is found on the Wrangell Island, a small island in the archipelago at the southeastern corner of Alaska. The site is different from the one in Leiknes as it is much closer to the shore. So close that part of it is submerged during high tide. But like the Norwegian counterpart, the site features many kinds of animals, most of which must have been important in hunting, like birds and salmon. The killer whale again is unmistakable. The head is big, with huge teeth, and the back features a tall triangular fin. It is decorated with figures on the body, maybe to illustrate the black-and-white markings on a killer whale, or maybe the figures have a symbolic meaning. It is one of those things that we will never know.

A thousand kilometers to the south of Wrangell Island, there is another rock carving that depicts two killer whales. The Wedding Rock Petroglyph Site is situated on the Olympic Peninsula in Washington State on the west coast of the United States. Like the site on Wrangell Island, there are many different carvings of both animals and human heads, or faces, and some geometric figures that look more like ancient doodles. The whales have open mouths and are characterized by their iconic tall erect back fins. The carvings even include stripes that could represent the white patch killer whales have on their sides. What ties together the art in Norway, on Wrangell Island, and on the Olympic Peninsula is that killer whales are still found in those areas today, as they must have been when the figures were made.

Places where you are close to ancient history have a way of getting to you. As we moved around to find all the figures on Dyreberget, we walked and talked quietly so as not to break the spell of the past. It was impossible not to feel the enchantment of the place. Even the children seemed to get it; they were quiet as well. My father found a place to sit after the challenging walk, and when he lit his pipe, I could tell not only how much he enjoyed being there, but also that he was taking in the place, century by century. The feeling of connection to whomever was here so many thousands of years ago was strong. The animals were so lively and fresh that they could have been made just days ago, and they were so obviously made by someone who shared our sense of aesthetics and love of wildlife. It was a humbling and overwhelming insight to feel how much we and the ancient artists were alike.

Whatever role killer whales played in these prehistoric people's lives and their ideas about nature and wildlife, it was clearly important enough that the artist behind the work put considerable time and energy into polishing the outline of this huge animal. Did they fear killer whales? Or treasure them? The famous biologist, writer, and thinker E. O. Wilson, who passed away in 2021, considered our relationship to predators in an article in *Discover* magazine: "*We are not just afraid of predators,*" he wrote, "*we're transfixed by them, prone to weave stories and fables and chatter endlessly about them, because fascination creates preparedness, and preparedness, survival. In a deeply tribal way, we love our monsters.*"[2]

Monsters come in many forms. Think of Leviathan or Moby Dick or the giant whale that swallowed Jonah. History and literature abound with legendary and mythological sea creatures wreaking havoc in their colossal monstrosity. And even when we understand that all these mythical creatures are just harmless whales, there are those who have maintained that killer whales, the biggest predators on earth, are dangerous to us or to our livelihoods, and the species has thus been persecuted with determination and brutality. It's important to note, however, that while there have been a number of killer whale attacks in marine parks that have ended with the death of a trainer or a caretaker, there are no verified fatal attacks on humans by killer whales in the wild.

But there have been encounters, or in some cases at least what have felt like attacks, leaving the people involved terrified and fearing for their lives. One such incidence affected British writer and journalist Philip Hoare, who was working with the Sri Lankan scientist Ranil P. Nanayakkara and his team off the coast of Sri Lanka. For more than an hour, a pod of eight killer whales had made attempts to attack a group of sperm whales with calves, but the sperm whales had successfully warded them off. Apparently frustrated with the failed hunt, the killer whales then turned their attention to Hoare and Nanayakkara's boat, circling closer and closer in a way that the people on board felt was aggressive, and at one point, even hitting it. "*Now we knew what it was like to be prey*," Hoare wrote in an article for the *Guardian*.[3]

To their relief, the killer whales disappeared for a bit, but it turned out it was only to regroup. They reemerged, lined up next to each other at the surface, from where they charged at full speed toward the boat in the same manner they have been seen attacking seals in Antarctica, by creating a wave that washes a seal off an ice floe. After this, Hoare and Nanayakkara fled the area at full speed, unwilling to test whether the killer whales were making mock charges or if it was serious business.

Recently, killer whales have stirred up trouble in Gibraltar and around the coasts of Spain and Portugal. The first incidences took place in 2020, when a small group of killer whales started interacting with sailboats and other small vessels, first in Gibraltar and then later along the Portuguese coast and in Galicia, in the northwestern corner of the Iberian Peninsula.

Imagine that you are in a sailboat, cruising along the coast in Gibraltar. Perhaps you just enjoyed lunch in the cockpit and are now relaxing in the sun. There is just enough wind to fill the sails and make the journey comfortable. Taking in the view, you spot black triangular fins approaching your boat, just before seeing the unmistakable black-and-white coloration of the large animals and realize the black fins belong to killer whales. Maybe you think it's a cool sight and you get your camera or your smartphone out, standing ready to film the whales as they get nearer. They come incredibly close—then disappear under the boat—and you think this is not just cool, it is awesome! Until the whole boat suddenly shudders as one of the whales rams the hull somewhere below.

You look over the railing in disbelief as the whales come up for breath, and when they dive down again and repeat the attack, it is no longer awesome—it is now incredibly frightening. And it goes on like this for hours, until the whales have chewed up the rudder and you can no longer maneuver the boat.

Incidents very much like this happened more than 200 times in 2020 and 2021. Since the first report in July 2020, 50 more were registered that year. Killer whales, at first only three individuals, would approach boats as they were sailing and purposely ram the hulls. They would also tug at rudders, sometimes hitting so hard the rudders would be abruptly snapped out of the hands of the helmsman.

In 2021, the number of episodes had more than tripled to 188 and, more worrying, the number of killer whales that participated in these interactions had also increased. Some sailors estimated that they had seen more than 30 whales participate in a single event. Not all "attacks" ended in damage to the boat, but approximately 50 of the interactions in 2021 ended with some sort of boat damage, typically to the rudder. Almost 40% of the cases involved serious damage to the boats, and quite a few had to be dragged to shore by another boat because they had become unsteerable.

Few people know these particular whales around Gibraltar and along the coasts of Spain and Portugal as well as Ruth Esteban, who is a marine mammal researcher with the Whale Museum in Madeira and has studied this population of killer whales for many years. In April 2022, I spoke with her by video call to get her thoughts on the situation.

She starts out by explaining that when a sailor spots killer whales in the water, the official recommendation is to immediately drop the sails or stop the engine and not engage with the animals at all. Esteban is a black-haired, soft-spoken woman, and behind her black glasses, her eyes smile when she admits that she really hasn't a clue why the killer whales in Gibraltar and adjacent areas have started wrecking boats.

"I could guess," she says, "but I am a scientist, and the fact is, I don't know."

Dropping the sails or turning the engine off discourages the whales—sometimes.

"It is not a foolproof thing, but so far it is the best advice we have," she says.

The killer whales in Gibraltar belong to a tiny population of just 39 that is critically endangered. They are specialists in catching tuna, sometimes actively hunting them, sometimes stealing them from the lines of fishermen, which is a strategy that often leads to conflict.[4] Since this small population of whales lives in some of the busiest and noisiest waters in the world, some people suggest that the whales' reaction to boats is a reaction to a stressful life, or even that they are retaliating against collisions with boats or conflicts with fishermen. Ruth Esteban is not so sure.

"It could be," she says, "but it could also just be that they, for some reason that we don't understand, find it entertaining; that it is a 'bad habit' they have developed and that something in our response, like yelling or screaming, has reinforced it." She is sure, however, that the behavior has spread and that many of the current population of those killer whales have now learned it from each other.[5]

Although it is entirely possible, maybe even plausible, that the whales are doing this for fun—the whale equivalent to a gang of roving hooligans, vandalizing just because they can—it doesn't make the behavior less frightening for those experiencing it. The interactions in Gibraltar are unexpected and difficult to explain, and we don't always get to decide how our relationship with a particular animal species evolves. With large predators like killer whales, wolves, lions, or bears, fear has always been a component in the relationship. A fear of predators is rational and protects us from harm—nobody in their right mind would walk straight up to a wild lion. But just as often, our relationships to predators are shrouded in misconceptions and myths.

The demonization of killer whales in earlier times has now been replaced by a different understanding. We know that killer whales breathe air like us but we also recognize in them many of our other behaviors. They take care of their young ones with as much dedication as we do, they grieve if a family member dies and may carry a dead calf for days or even weeks after its death, they frolic and surf in the waves and throw jellyfish as if they are frisbees, and they communicate like us, too. They are not monsters—not even when they are wrecking boats. Today, killer whales are the pride of many local communities where they were previously pursued, and through wildlife tourism and whale watching, the whales contribute enormously to the local economies. Killer whales have become the icons of whole regions and are used as brands by the tourism industry, as well as on merchandise and to market local products.

This shift in perception and attitude can be surprisingly swift, as I have witnessed in Norway. In Iceland, as in Norway, killer whales have also become the mascots of the country, although they must compete with the island's famously furry horses and cute puffins. In North America, the shift took place a few decades earlier, but there, too, it was a quick change from shooting at killer whales with machine guns to a state of appreciation and devotion. These shifts in attitude give us hope that places where killer whales are still persecuted, hunted, or captured to be traded for display in oceanariums and marine parks will change their ways. It is possible to reverse the situation, even to the benefit of those who have felt their livelihoods threatened.

It is not difficult to parse the motivations for this change. Knowledge plays a key part. We appreciate killer whales as unique animals. We admire their intelligence and see part of our own behavior reflected in another species. We connect with them. And the more we learn about killer whales' behavior and social lives, the more strongly we connect with them. Meeting a large predator like a killer whale is perhaps one of the most potent ways of connecting with nature. The immensity, the beauty, and the power are overwhelming. They are immersed in an element that is not ours, and when we connect to them, we can, for a fleeting second, cross the border to a realm that is usually out of our reach.

Whoever drew the animals on the rocky slopes at Dyreberget must have felt a longing to connect, too. The tenderness in the artist's rendering of the animals, the detail in the portrayal of the different species' characteristics, and the careful observation of their movements makes it evident that he or she had intimate knowledge of these animals. Perhaps they revered the moose, the reindeer, the swans, and the killer whale, whose outlines were so carefully polished on the rocks. Most likely they also chased them, or at least some of them. Living this far north, survival must have been an ever-present challenge, in which every catch of any species could make the difference between life and death.

Archaeologists don't know why these figures were made or what they meant to the people who made them, but it is generally believed that the species pictured there and at other sites played a role as game animals. They were species hunted by the prehistoric people living here, and picturing them on the rocks could have been part of a religious or ritualistic ceremony. But I wonder about the killer whale. Was that a game species, too? Or was it placed here with the other animals because it would come along with other species that they hunted or fished, like herring or saithe or seals? Or did it bring other whales to the shore as people in different parts of the world have noticed, and in some places even learned to benefit from? There are petroglyphs in Norway picturing whaling scenes, with people in boats apparently chasing whales and attaching lines to them. But at Dyreberget there are only figures of animals, none of people, so these questions are left unanswered.

It was already late when we had started our climb up to the site, and while we explored, the light faded quickly. The few clouds in the deep blue sky turned from pink to purple and the first stars had risen. We needed to hurry back if we didn't want to descend in total darkness. We walked slowly and carefully, but still my father stumbled and we could hear him swearing every time he was thrown off balance or his feet got caught on roots running across the path. The children took his hands and led him, but the stumbling and cursing continued until somehow we all managed to make it down to the boggy area. It was darker there, but the path was more even. Soon we were back on the road and to the car, giggling with relief.

Meeting killer whales and the people associated with them changed the trajectory of my life. There is always a certain amount of chance involved in life and in the decisions we make, and I let that element guide me when I decided to become a cook on a small rebuilt fishing vessel many years ago. But it wasn't chance that made me stay. It was the opportunity to be surrounded by wilderness, to be close to wildlife, and to be with like-minded people. Life in all its variety and beauty is endlessly fascinating, and I am quite sure I could have been equally captivated by the lives of butterflies or gorillas or pine trees or even herring. But that chance conversation while waiting in line at the university cafeteria led me to killer whales.

When I stopped working at the Whale Center in the late 1990s and moved back to Denmark, I thought I had given up the whales for the security of a steady and well-paying job. Then in 2016, when the opportunity arose to return to Norway with a research group and a film team, it felt like a soothing bandage on an open wound to be back at sea. Encountering the whales again was as exhilarating as it was when I had first seen them.

One night, as the field season that year was coming to an end, I was driving back to the small cottages where the film team and I were staying. I had been in Tromsø to talk to Professor Audun Rikardsen about the relationship between fishermen and killer whales, and my return trip wove across the mountains in the dark winter landscape. Outside Tromsø there was still a lot of traffic, but after a few hours driving, the traffic thinned and the roads were empty. It was a moonless night. Occasionally, green bands of northern lights waved across the sky, and a few times, a snow hare jumped in front of the car, making me hit the brakes—which was unnecessary since the hares were always so fast they were far away across a snow field by the time I'd react.

I can't always pinpoint precisely when I've made a crucial decision in life, or what makes it tip in one direction instead of another. I certainly don't know why exactly I made my next big life decision during that drive back from Tromsø, and I don't recall any lengthy back-and-forth in my head to justify it. I just know that when I arrived at the cottages, I had made a decision. Perhaps I was again guided by the philosophy of my childhood heroine Pippi Longstocking: "*If you are very strong, you must also be very kind.*" She probably didn't have the environment in mind when she made that determination, but I have chosen to interpret this as a reminder that when it comes to nature, we are the strong ones; it is not the fish or the birds, not the trees or the animals—it is not even the killer whales, but us. We are the ones in power, and it is up to us to be kind, to make a better world for all of us living creatures who share this planet together.

The others were still up when I arrived at the cottages, and as they helped unload the car, I slipped inside to formalize my decision to return to a life more connected to the sea. It felt like an echo of the decision I had made years earlier, when I put my college studies on pause to join that first killer whale expedition. I hit send on an email to the director of the Natural History Museum of Denmark resigning my position. Then I went to join the others for a beer.

Acknowledgments

For introducing me to the world of whales and the ways of the sea, and for putting up with my cooking, I am eternally grateful to Mic Calvert, Bibi Forsman, Morten Lindhard, and all the others who were onboard *Old-Bi* during the early years working in the Lofoten Islands. Special thanks go to the Center for Studies of Whales and Dolphins, the Gothenburg-based nongovernmental organization that initiated the project *Old-Bi* was part of, and which eventually led to the establishment of the Whale Center in Andenes, Norway.

On board *Old-Bi* and later at the Whale Center, I met so many people whose knowledge, curiosity, and sharp minds created an intellectually challenging playground and gave me the inspiration to dive into the biology of whales and animal behavior. They are listed alphabetically: Arnim Andreae, Susanne Bang, Martin Biuw, Sonja Blom, Jenny Burdon, Stefan Burger, Turid Buvik, Mads Fage Christoffersen, Jacki Ciano, Catherine DeNardo, Michael Dougherty, Göran Ehlmé, Allan Falk,

Vivi Fleming, Peter van der Gulik, Jan Herrmann, Bo Johanneson, Roar Jørgensen, Lars Øivind Knutsen, Erland Lettevall, Thomas Lyrholm, Laurence Mazaudier , Lise-Lotte Medlund, Anne Kamp Moth-Poulsen, Leif Nøttestad, Annika Rockström, Tuula Rotkokivi, Tiu Similä, Jonas Teilmann, Graham Timmins, Fernando Ugarte, and Magnus Wahlberg. Undoubtedly there are some whose names have slipped my mind; I extend my heartfelt thanks to them, as well. Many people in and around Andenes also contributed to making the work at the Whale Center enjoyable and rewarding. Special thanks go to Erwin Fulterer, Atle Hagtun, Camilla Ilmoni, Åge Ingebrigtsen, Helge Ingebrigtsen, Geir Maan, Glenn Maan, Svein Spjelkavik, and Daniele Zanoni.

Working in the field often involves hard work and frustrations with foul weather or failing equipment, but when the season ends those challenges tend to fade while the memories of encounters with wild animals and the wonder of being surrounded by wilderness endure. To me, sharing these moments with others is the high point of fieldwork. I would like to extend my deepest gratitude for all these shared moments to Martin Biuw, Johan Candert, Tobias Dahlin, Catherine DeNardo, Rune Dietz, Michael Dougherty, Göran Ehlmé, John and Beverly Ford, Catharina Frostad, Martin Halvorsen, Mikkel Villum Jensen, Anders Drud Jordan, Lars Øivind Knutsen, Henrik Egede Lassen, William Long, Thomas Lyrholm, Per Ole Lund, Christian Mailand, Cristina Mittermeier, Anne Kamp Moth-Poulsen, Birgitta Mueck, Paul Nicklen, Øivind Nielssen, Christina Lehmkuhl Noer, Joakim Odelberg, Morten Tange Olsen, Audun Rikardsen, Tiu Similä, Simon Stanford, Frederik Wolff Teglhus, Fernando Ugarte, and Magnus Wahlberg.

While conducting research for this book, I received help in many forms and from many people. For valuable discussions and for assistance in finding the right literature and the right contacts, my sincere thanks go to Line Christiansen from Sermilik Adventures in Greenland; Roslyn Heather Davidson and Olga Filatova from Moscow State

University; Abdi Hedayat from the Natural History Museum of Denmark; Erich Hoyt, Carl Christian Kinze, Tiu Similä, and Kristjana Vigdís from the National Archive of Iceland; Angela George and Jody White from the Killer Whale Museum in Eden, Australia; and Mads Peter Heide-Jørgensen and Fernando Ugarte from the Greenland Institute of Natural Resources.

In Kamchatka, Russia, Tanya Ivkovich and her team in the camp at Zelyony Mys, the Green Cape, met me with incredible hospitality. My sincere thanks for great conversations, close encounters with killer whales (and poachers), good food, and hot showers go to Asya Danishevskaya, Alexandra Dombrovskaya, Ivan Nikolayev, Evgenii Smirnov, Dmitry Voronov, Julia Smirnova, and especially to Tanya, who gave me invaluable insights into the biology and conservation of the killer whales in Kamchatka.

I am incredibly grateful for the generous help I received from Tobias Ignatiussen and Gedion Ignatiussen in Tasiilaq, Greenland, including their willingness to discuss the sensitive subject of hunting and to share their knowledge of the area and its wildlife. I also want to thank Robert Peroni from The Red House for his hospitality and for lending me books, keeping me company, offering home-cooked dinners, and sharing his love of Greenland and the Inuit way of life. Thanks also to Hanne Bech Danielsen for great walks and talks.

In Norway, when not living on a boat, I have been offered the comforts of a warm bed, cozy dinners, and good company by Per Ole Lund, Anna-Lisa Mitrovic, and Trym Ivar Bergsmo and Runa Bergsmo. Thank you.

I am deeply indebted to those who offered their stories, expertise, opinions, and experience in conversations and interviews with me. You have all contributed greatly to shedding more light on our sometimes troubled relationship with killer whales: Sigrund Krane, Austnesfjord, Norway; David Hancock, the Hancock Wildlife Foundation in Canada; Masha Netrebenko, Vesti Primorye (the Kamchatka regional branch of the Russian State TV); Rune Dietz and Christian Sonne, Aarhus University, Denmark; Ernest Alfred, Hereditary Chief of the Tlowit'sis people and elected councilor of the 'Namgis First Nation; Eve Jourdain, who studies killer whales in Norway; and Ruth Esteban, the Whale Museum in Madeira. A special thanks goes to Marco Shulenburg, whose intimate knowledge of killer whales, gained through many years of carefully following in their wake, was very helpful when I started doing research for this book.

I am grateful for the help translating texts from Icelandic to either Danish or English that I have received from Jesper Lauridsen and Torsteinn Helgasson. In Greenland, a young woman whose name I didn't get helped me communicate with the Greenlandic-speaking Inuits. My thanks to her, as well.

I would like to thank Astrid van Ginneken for permission to quote her carefully written field notes from 1994, the Australian Broadcasting Company for permission to quote from interviews in the 1998 documentary *Killers in Eden* and from an interview with Guboo Ted Thomas in 2014, and Dr. Robert Pitman and Dr. John W. Durban for permission to quote from their 2009 article in *Natural History Magazine*.

I would like to extend my sincere thanks to the scientific and other expert sources in the field who have read sections or chapters of the book and offered their comments and suggestions: Ernest Alfred, Catherine DeNardo, Rune Dietz, Göran Ehlmé, Ruth Esteban, Tanya Ivkovich, Lars Øivind Knutsen, Audun Rikardsen, Christian Sonne, and Fernando Ugarte. Also sincere thanks to Camilla Lindhard Karas, Henrik Egede Lassen, and Graham Timmins; they have all done a great job commenting on all aspects of the book.

The writings of Bernd Heinrich have been a profound inspiration to me since I read his wonderful book *Ravens in Winter* many years ago. I am therefore particularly grateful for the extremely kind and helpful suggestions I received from Bernd when I started looking for a literary agent. And I do have a fantastic agent! My sincerest thanks go to my agent Elise Capron, who believed in this book and helped me strengthen the concept and the story.

My sincere thanks to science journalist, photographer, and editor Emily Harwitz and to Senior Acquisitions Editor Tiffany Gasbarrini at Johns Hopkins University Press, who have been incredibly helpful in their suggestions for improving the book. Both have contributed with overall ideas and concepts as well as more specific recommendations for changes. It was a pleasure working closely with Emily, whose suggestions for adjustments, clarifications, additions, and reorganizations made the book better at so many levels. Joanne Haines did so much more than copyediting. She caught the billions of small and large errors, she offered heaps of suggestions for improvement in both writing and content, and she went the extra mile to fact-check information.

The book would not have been what it is—in fact, it wouldn't have been at all—if not for the help and encouragement I have received from Catherine DeNardo, who is a marine mammal colleague, a fellow writer, and a dear friend. Thank you for the enthusiasm and commitment as well as for all the discussions, comments, and suggestions. More than anything, your involvement made writing the book more fun.

Lastly, I would like to extend my deepest gratitude to Paul Nicklen, not only for writing a beautiful foreword for the book and offering his incredible pictures for it but for being a wonderful friend—and for using his voice and photographs to create awareness and love for everything wild, including killer whales.

Notes

CHAPTER 1

Bloody Beasts

1. Pliny the Elder 1855, 365–66.
2. Larson 1917, 119–120.
3. Eschricht 1862.
4. Barrett-Lennard et al. 1995.
5. Thestrup 1999.
6. Verne 1876, 262.
7. Hedayat 2015.

CHAPTER 2

Sea Change

1. Katona et al. 1979.
2. Ford, Ellis, and Balcomb 1996.
3. Barrett-Lennard 2000.
4. Ford and Ellis 1999.
5. Ford 1991.
6. Whitehead and Rendell 2014.
7. Strager 1995.
8. Astrid van Ginneken, unpublished field notes, 1993.
9. Lorenz 1961.
10. Andersen 2007, 121.

CHAPTER 3

Blubber Choppers

1. Robertson 1954.
2. Robertson 1954, 158.
3. Whitehead and Reeves 2005.
4. Andrews 1916, 201.
5. Andrews 1916, 221.
6. Andrews 1916, 200–201.
7. Utne 1932, 34.
8. Utne 1932, 34.

CHAPTER 4

The Law of the Tongue

1. Mathews 1904.
2. Morgan 1914, 9.
3. Dakin 1934, as quoted in Whitehead and Rendell 2014, 141–42.
4. Ireland 1998.
5. Blaxter 1998.
6. Clode 2002, 88.
7. Stead 1930, 9.
8. Otton 1998.
9. *Twofold Bay Magnet* 1912, 4.
10. *Australian Town and Country Journal* 1906, 28.
11. *Australian Town and Country Journal* 1894, 12.
12. Kingsley 1899.
13. *Illawarra Mercury*, 1875, 4.
14. Stead 1930, 9.
15. *Eden Observer and South Coast Advocate*, August 3, 1906.
16. Simões-Lopes, Fabián, and Menegheti 1998.
17. Brown 2014, Guboo Ted Thomas starts his story at 0:7:00.
18. Gibbs 2010.
19. *Daily Telegraph* 1906, 18.
20. Stead 1930, 9.

CHAPTER 5

War Zone

1. *White Falcon Newsletter*, various issues, 1954–1957, US Naval Station, Keflavik Airport, Iceland.
2. *Time* 1954.
3. *White Falcon Newsletter* 1955, 1.
4. Martinussen 2006.
5. Christensen 2014.
6. Christensen 2014.
7. Øien 1988.
8. Harding et al. 2020.
9. Sunde et al. 2021.
10. Dahlstöm 2009.
11. Bowen and Lidgard 2012.
12. Goodall 1990, 165.
13. David Hancock, unpublished memoirs, 2020.
14. Colby 2018.
15. *Douglas Island News* 1920, 2.

CHAPTER 6

A Turn for the Better

1. Larson 1917, 120–21.
2. Larson 1917, 121.
3. Larson 1917, 121.
4. OECD and Sandberg 2010.
5. Similä and Ugarte 2011.

CHAPTER 7

The Whales in the Potato Field

1. Haug and Sandnes 1982.
2. Darwin 1871, 101.
3. Miralles, Raymond, and Lecointre 2019.
4. Pitman et al. 2017.
5. Pitman and Durban 2009, 48.

CHAPTER 8

The Whales at the End of the World

1. Steller 1988, 115.
2. Steller 2011, 79.
3. Levin 2010, 0:12:05 to 0:12:15.

CHAPTER 10

When the Hunters Become the Hunted

1. StatBank Greenland n.d.
2. Rosing-Asvid and Ugarte 2013.
3. Whitehead and Rendell 2014.
4. Sargeant, Forsyth, and Pitman 2018.
5. Schmidt 2020.
6. MacKenzie et al. 2014.
7. Ugarte et al. 2021.
8. Ugarte et al. 2021.
9. Ugarte et al. 2021.
10. Ugarte et al. 2021.
11. Sommer and Lihn 2018, para. 5.
12. Desforges et al. 2018.
13. Ugarte et al. 2021.
14. Yodzis 2001.
15. Rasmussen 2005.
16. Jourdain et al. 2020.

CHAPTER 11

Family Matters

1. Andersen 2002.
2. *Olympian* 2002, B5.
3. Cederholm et al. 2001.
4. Governor's Salmon Recovery Office 2021.
5. Ford and Myers 2008.
6. Ford and Ellis 2006.
7. Ford et al. 2010.
8. Groskreutz et al. 2019.
9. Towers et al. 2020.
10. Governor's Salmon Recovery Office 2021.
11. Department of Fisheries and Oceans 2018.

CHAPTER 12

Cut in Stone

1. Helberg 2016.
2. Wilson 1985.
3. Hoare 2017.
4. Esteban et al. 2016.
5. Esteban et al. 2022.

Bibliography

Andersen, Flavio Duarte, 2007. *Norsk hvalfangstpolitikk 1972–1982: opinionen for og opposisjonen mot hvalfangst i perioden 1972–1982* [Norwegian whaling politics 1972–1982: The opinion for and the opposition against whaling in the period 1972–1982]. Master's thesis, University of Oslo.

Andersen, Peggy. 2002. Orphaned orca released to join pod. *Associated Press*. July 15.

Andrews, Roy Chapman. 1916. *Whale Hunting with Gun and Camera*. New York: D. Appleton.

Australian Town and Country Journal. 1894. The blubber cure. November 24.

Australian Town and Country Journal. 1906. The whalers of Twofold Bay. August 29.

Barrett-Lennard, Lance G. 2000. *Population structure and mating patterns of killer whales (Orcinus orca) as revealed by DNA analysis*. PhD diss., University of British Columbia.

Barrett-Lennard, Lance G., Kathy Heise, Eva Saulitis, Graeme Ellis, and Craig Matkin. 1995. *The Impact of Killer Whale Predation on Steller Sea Lion Populations in British Columbia and Alaska*. Report for the North Pacific

Universities Marine Mammal Research Consortium, University of British Columbia, Vancouver, BC, Canada.

Blaxter, Bill. 1998. Quote from an interview in the ABC Documentary *Killers in Eden*, 1998. With permission from the Australian Broadcasting Corporation.

Bowen, William, and Damian Lidgard. 2012. Marine mammal culling programs: Review of effects on predator and prey populations. *Mammal Review* 43 (3): 207–20. https://doi.org/10.1111/j.1365-2907.2012.00217.x.

Brown, Bill. 2014. The Aboriginal whalers of Eden. *ABC Local*. https://www.abc.net.au/local/audio/2013/10/29/3879462.htm?site=southeastnsw.

Carson, Rachel L. 1962. *Silent Spring*. Boston: Houghton Mifflin.

Cederholm, C. Jeff, David H. Johnson, Robert E. Bilby, Lawrence G. Dominguez, Ann M. Garrett, William H. Graeber, Eva L. Greda, et al. 2001. Pacific salmon and wildlife: Ecological contexts, relationships, and implications for management. In *Wildlife-Habitat Relationships in Oregon and Washington*, edited by Thomas A. O'Neil and David H. Johnson, 628–85. Corvallis: Oregon State University Press.

Christensen, Pål. 2014. Silda: fra kanonfiske til katastrofe [Herring: From boundless fishing to catastrophe]. In Norges fiskeri- og kysthistorie, bind IV: havet, fisken, og oljen, 1970–2014 [Norway's fishery and coastal history, volume IV: Sea, fish, and oil, 1970–2014]. Bergen: Fagbokforlaget.

Clode, Danielle. 2002. *Killers in Eden: The True Story of Killer Whales and Their Remarkable Partnership with the Whalers of Twofold Bay*. Crows Nest, New South Wales: Allen & Unwin.

Colby, Jason M. 2018. *Orca: How We Came to Know and Love the Ocean's Greatest Predator*. New York: Oxford University Press.

Dahlström, Åsa Nilsson. 2009. "Shoot, dig, and shut up!" Differing perceptions of wolves in urban and rural Sweden. *Ethnologie Française* 39 (1): 101–8.

Daily Telegraph. 1906. Protection for the "killer." October 27.

Dakin, William John. 1934. *Whalemen Adventurers: The Story of Whaling in Australian Waters and Other Southern Seas Related Thereto, from the Days of Sails to Modern Times*. Sydney: Angus & Robertson.

Darwin, Charles. 1871. *The Descent of Man, and Selection in Relation to Sex*. London: John Murray.

Department of Fisheries and Oceans. 2018. *Recovery Strategy for the Northern and Southern Resident Killer Whales (Orcinus orca) in Canada* [Proposed]. Species at Risk Act Recovery Strategy Report Series. Ottawa: Fisheries and Oceans Canada.

Desforges, Jean-Pierre, Ailsa Hall, Bernie McConnell, Aqqalu Rosing-Asvid, Jonathan L. Barber, Andrew Brownlow, et al. 2018. Predicting global killer whale population collapse from PCB pollution. *Science* 361 (6409): 1373–76.

Douglas Island News. 1920. The eagle. August 6.

Eschricht, Daniel Friedrich. 1862. Om spækhuggeren (Delphinus Orca L.) [About the killer whale (Delphinus Orca L.)]. Lecture held at the Royal Danish Academy of Sciences and Letters, May 9, 1862.

Esteban, Ruth, Alfredo López, Álvaro de los Rios, Marisa Ferreira, Francisco Martinho, Paula Méndez-Fernandeza, Ezequiel Andréu, José C. García-Gómez, Liliana Olaya-Ponzone, Rocío Espada-Ruiz, Francisco J. Gil-Vera, Cristina Martín Bernal, Elvira Garcia-Bellido Capdevila, Marina Sequeira, and José Martínez-Cedeira. 2022. Killer whales of the Strait of Gibraltar, an endangered subpopulation showing a disruptive behavior. *Marine Mammal Science*, June 8, 2022. https://doi.org/10.1111/mms.12947.

Esteban, Ruth, Philippe Verborgh, Pauline Gauffier, Joan Giménez, Christophe Guinet, and Renaud de Stephanis. 2016. Dynamics of killer whale, bluefin tuna and human fisheries in the Strait of Gibraltar. *Biological Conservation* 194 (February): 31–38.

Ford, Jennifer S., and Ransom A. Myers. 2008. A global assessment of salmon aquaculture impacts on wild salmonids. *PLoS Biology* 6 (2): e33. https://doi.org/10.1371/journal.pbio.0060033.

Ford, John K. B. 1991. Vocal traditions among resident killer whales (*Orcinus orca*) in coastal waters of British Columbia. *Canadian Journal of Zoology* 69 (6): 1454–83.

Ford, John K. B., and Graeme M. Ellis. 1999. Transients: mammal-hunting killer whales of British Columbia, Washington, and southeastern Alaska. Vancouver: University of British Columbia Press.

Ford, John K. B., and Graeme M. Ellis. 2006. Selective foraging by fish-eating killer whales *Orcinus orca* in British Columbia. *Marine Ecology Progress Series* 316:185–99.

Ford, John K. B., Graeme M. Ellis, and Kenneth C. Balcomb. 1996. Killer whales: the natural history and genealogy of *Orcinus orca* in British Columbia and Washington. Vancouver: University of British Columbia Press.

Ford, John K. B., Graeme M. Ellis, Peter F. Olesiuk, and Kenneth C. Balcomb. 2010. Linking killer whale survival and prey abundance: Food limitation in the oceans' apex predator? *Biology Letters* 6 (1): 139–42.

Gibbs, Martin. 2010. *The Shore Whalers of Western Australia: Historical Archaeology of a Maritime Frontier*. Australia: Sydney University Press.

Goodall, Jane. 1990. *Through a Window*. London: Weidenfeld and Nicolson.

Governor's Salmon Recovery Office. 2021. *2020 State of Salmon in Watersheds.* Olympia: Washington State Recreation and Conservation Office. https://stateofsalmon.wa.gov.

Groskreutz, Molly J., John W. Durban, Holly Fearnbach, Lance G. Barrett-Lennard, Jared R. Towers, and John K. B. Ford. 2019. Decadal changes in adult size of salmon-eating killer whales in the eastern North Pacific. *Endangered Species Research* 40:183–88.

Harding, Lee E., Mathieu Bourbonnais, Andrew T. Cook, Toby Spribille, Viktoria Wagner, and Chris Darimont. 2020. No statistical support for wolf control and maternal penning as conservation measures for endangered mountain caribou. *Biodiversity and Conservation* 29 (9): 3051–360.

Haug, Tore, and Otto Kristian Sandnes. 1982. Massestranding og vellykket berging av spekkhoggere i Lofoten [Mass stranding and successful rescue of killer whales in Lofoten]. *Fauna* 35:1–7.

Hedayat, Abdi. 2015. Døde dyr fra København i verdenslitteraturen [Dead animals from Copenhagen in world literature]. Naturhistorier, nr. 1, 26–28. Statens Naturhistoriske Museum.

Helberg, Bjørn Hebba. 2016. Bergkunst nord for polarsirkelen [Rock art north of the Arctic Circle]. Tromsø Museums Skrifter.

Hoare, Philip. 2017. An extraordinary battle between sperm whales and orcas—in pictures. *Guardian*, March 27, 2017.

Hoyt, Erich. 2013. *Orca: The Whale Called Killer*. Richmond Hill, ON: Firefly Books.

Illawarra Mercury. 1875. Whale captured at Twofold Bay. September 3.

Ireland, Douglas. 1998. Quote from an interview in the ABC documentary

Killers in Eden, 1998. With permission from the Australian Broadcasting Corporation.

Jourdain, Eve, Clare Andvik, Richard Karoliussen, Anders Ruus, Dag Vongraven, and Katrine Borgå. 2020. Isotopic niche differs between seal and fish-eating killer whales (*Orcinus orca*) in northern Norway. *Ecology and Evolution* 10 (9): 4115–27. https://doi.org/10.1002/ece3.6182.

Katona, Steven, Ben Baxter, Oliver Brazier, Scott Kraus, Judy Perkins, and Hal Whitehead. 1979. Identification of humpback whales by fluke photographs. In *Behavior of Marine Animals*, edited by Howard E. Winn and Bori L. Olla, 33–44. Boston: Springer. https://doi.org/10.1007/978-1-4684-2985-5_2.

Kingsley, Mary Henrietta. 1899. *West African Studies*. London: Macmillan.

Larson, Laurence Marcellus, trans. 1917. *The King's Mirror* [Speculum regale]. New York: American-Scandinavian Foundation.

Levin, Daniel. 2010. *Kamchatka: The Salmon Country*. Good Hope Films. 20 minutes. https://www.imdb.com/title/tt1615462/.

Lorenz, Konrad. 1961. *King Solomon's Ring*. Translated by Marjorie Kerr Wilson. London: Methuen.

MacKenzie, Brian R., Mark R. Payne, Jesper Boje, Jacob L. Høyer, and Helle Siegstad. 2014. A cascade of warming impacts brings bluefin tuna to Greenland waters. *Global Change Biology* 20 (8): 2484–91. https://doi.org/10.1111/gcb.12597.

Martinussen, A. O. 2006. Nylon fever: Technological innovation, diffusion and control in Norwegian fisheries during the 1950s. *Maritime Studies* 5 (1): 29–44.

Mathews, Robert Hamilton. 1904. Ethnological notes on the Aboriginal tribes of New South Wales and Victoria. *Journal & Proceedings of the Royal Society of New South Wales* 38: 203–381.

Miralles, Aurélien, Michel Raymond, and Guillaume Lecointre. 2019. Empathy and compassion toward other species decrease with evolutionary divergence time. *Scientific Reports* 9:19555. https://doi.org/10.1038/s41598-019-56006-9.

Morgan, James. 1914. The killers of Twofold Bay. *Sydney Mail*. June 17.

National Archive of Iceland. 1950–1959. Various newspaper articles.

Neiwert, David. 2015. *Of Orcas and Men: What Killer Whales Can Teach Us.* New York: Abrams.

Obee, Bruce, and Graeme Ellis. 1992. *Guardians of the Whales: The Quest to Study Whales in the Wild.* Vancouver, BC: Whitecap Books.

OECD (Organisation for Economic Co-operation and Development) and Per Sandberg. 2010. Rebuilding the stock of Norwegian spring spawning herring: Lessons learned. In *The Economics of Rebuilding Fisheries: Workshop Proceedings*, 219–33. Paris: OECD Publishing. https://doi.org/10.1787/9789264075429-12-en.

Olympian. 2002. Orphaned orca facing first winter at sea after Puget Sound Rescue. November 13.

Otton, Alice. 1998. Quote from an interview in the ABC documentary *Killers in Eden*, 1998. With permission from the Australian Broadcasting Corporation.

Pitman, Robert L., Volker B. Deecke, Christine M. Gabriele, Mridula Srinivasan, Nancy Black, Judith Denkinger, John W. Durban, et al. 2017. Humpback whales interfering when mammal-eating killer whales attack other species: Mobbing behavior and interspecific altruism? *Marine Mammal Science* 33 (1): 7–58. https://doi.org/10.1111/mms.12343.

Pitman, Robert L., and John W. Durban. 2009. Save the seal! *Natural History Magazine* November 2009:48.

Pliny the Elder. 1855. *The Natural History*. Edited by John Bostock. London: Taylor and Francis. http://www.perseus.tufts.edu/hopper/text?doc=urn:cts:latinLit:phi0978.phi001.perseus-eng1.

Rasmussen, Henriette. 2005. Sustainable Greenland and indigenous ideals. In *The Earth Charter in Action: Toward a Sustainable World*, edited by Peter Blaze Corcoran, 106–8. Amsterdam: Royal Tropical Institute. https://earthcharter.org/wp-content/assets/virtual-library2/images/uploads/ENG-Rasmussen.pdf.

Robertson, Robert Blackwood. 1954. *Of Whales and Men*. New York: Simon & Schuster.

Rosing-Asvid, A., and F. Ugarte. 2013. Notat om spækhuggerfangst i Grønland [Note from Greenland Institute of Natural Resources to Ministry of Fisheries and Hunting, Greenland]. September 25.

Sargeant, Hannah, Rebecca Forsyth, and Alexandra Pitman. 2018. The epidemiology of suicide in young men in Greenland: A systematic review. *International Journal of Environmental Research and Public Health* 15 (11): 2442. https://doi.org/10.3390/ijerph15112442.

Schmidt, Gudrun M. 2020. Interview with Niviaq Korneliussen. *Politiken*, August 22.

Similä, Tiu, and Fernando Ugarte. 2011. Surface and underwater observations of cooperatively feeding killer whales in northern Norway. *Canadian Journal of Zoology* 71 (8): 1494–99. https://doi.org/10.1139/z93-210.

Simões-Lopes, Paulo, Marta Fabián, and João Menegheti. 1998. Dolphin interactions with the mullet artisanal fishing on Southern Brazil: A qualitative and quantitative approach. *Revista Brasileira de Zoologia* 15 (3): 709–26. https://doi.org/10.1590/S0101-81751998000300016.

Sommer, Karsten, and Anton Gundersen Lihn. Råd advarer mod at spise spækhuggere [Counsel warns against eating killer whale meat]. KNR: Greenlandic Broadcasting Corporation. https://knr.gl/da/nyheder/råd-advarer-mod-spise-spækhugger.

StatBank Greenland. n.d. Catches of mammals and birds, Greenland by species and time, 2006–2020. Accessed May 21, 2022. https://bank.stat.gl/pxweb/en/Greenland/Greenland__FI__FI20/FIXFANGST.px/table/tableViewLayout1.

Stead, David G. 1930. Tom, the killer, and his friends. *Sydney Morning Herald*. October 4.

Steller, Georg Wilhelm. 1988. *Journal of a Voyage with Bering, 1741–1742*. Translated by M. A. Engel and O. W. Frost. Edited by O. W. Frost. Stanford: Stanford University Press.

Steller, Georg Wilhelm. 2011. *De bestiis marinis, or The Beasts of the Sea*. Translated by Walter Miller and Jennie Emerson Miller. Edited by Paul Royster. Lincoln, NE: Zea E-Books. https://digitalcommons.unl.edu/zeabook/1/.

Strager, Hanne. 1995. Pod-specific call repertoires and compound calls of killer whales, *Orcinus orca* Linnaeus, 1758, in the waters of northern Norway. *Canadian Journal of Zoology* 73 (6): 1037–47.

Sunde, Peter, Sebastian Collet, Carsten Nowak, Philip Francis Thomsen, Michael Møller Hansen, Björn Schulz, Jens Matzen, Frank-Uwe Michler,

Christina Vedel-Smith, and Kent Olsen. 2021. Where have all the young wolves gone? Traffic and cryptic mortality create a wolf population sink in Denmark and northernmost Germany. *Conservation Letters* 14 (5): e12812.

Thestrup, Poul. 1999. Mark og skilling, kroner og øre: pengeenheder, priser og lønninger i Danmark i 360 år (1640–1999) [Mark, shilling, and crowns: monetary units, prices and salaries in Denmark in 360 years (1640–1999)]. Statens Arkiver.

Time. 1954. Iceland: Killing the killers. October 4. https://content.time.com/time/subscriber/article/0,33009,857557,00.html.

Towers, Jared R., James Pilkington, Brian Gisborne, Brianna Wright, Graeme M. Ellis, John K. B. Ford, and Thomas Doniol-Valcroze. 2020. Photo-identification catalogue and status of the northern resident killer whale population in 2019. *Canadian Technical Report of Fisheries and Aquatic Sciences* 3371: iv + 69 pp.

Twofold Bay Magnet and South Coast and Southern Monaro Advertiser. 1912. The killers. July 8.

Ugarte, Fernando, Mads Peter Heide-Jørgensen, Kristin Laidre, and Aqqalu Rosing-Asvid. 2021. Local knowledge about killer whales in narwhal grounds of West and East Greenland. Working paper NAMMCO SC/28/NEGWG/17. Presented at the Ad hoc Working Group on Narwhal in East Greenland, October 25–29, 2021, Copenhangen, Denmark.

Utne, Martha Brock. 1932. To jordebøker fra 1694 [Two land registers from 1694]. In Nordnorske Samlinger utgitt av Etnografisk Museum i Finnmark, Første hefte [North Norwegian Collections, published by the Ethnographic Museum in Finnmark, first booklet]. Oslo: A.W. Brøggers Boktrykkeri A/S.

Verne, Jules. 1875. *Twenty Thousand Leagues Under the Seas*. Boston: J. R. Osgood.

White Falcon Newsletter. 1955. VP18 patrol planes attack killer whales. October 29.

Whitehead, Hal, and Randall Reeves. 2005. Killer whales and whaling: The scavenging hypothesis. *Biology Letters* 1 (4): 415–8. https://doi.org/10.1098/rsbl.2005.0348.

Whitehead, Hal, and Luke Rendell. 2014. *The Cultural Lives of Whales and Dolphins*. Chicago: University of Chicago Press.

Wilson, Edward Osborne. 1985. In praise of sharks. *Discover* 6 (7): 40–53.

Yodzis, Peter. 2001. Must top predators be culled for the sake of fisheries? *Trends in Ecology & Evolution* 16 (2): 78–84. https://doi.org/10.1016/S0169-5347(00)02062-0.

Øien, N. 1988. The distribution of killer whales (*Orcinus orca*) in the North Atlantic based on Norwegian catches, 1938–1981, and incidental sightings, 1967–1987. *Rit Fiskideildar* 11:65–78.

Index